CDPD

McGraw-Hill Series on Computer Communications

CDPD

**Cellular Digital Packet Data
Standards and Technology**

John Agosta

Travis Russell

McGraw-Hill

New York   San Francisco   Washington, D.C.   Auckland   Bogotá
Caracas   Lisbon   London   Madrid   Mexico City   Milan
Montreal   New Delhi   San Juan   Singapore
Sydney   Tokyo   Toronto

Library of Congress Cataloging-in-Publication Data

Agosta, John.
    CDPD : cellular digital packet data standards and technology /
John Agosta, Travis Russell.
        p.   cm.—(McGraw-Hill series on computer communications)
    Includes index.
    ISBN 0-07-000600-8
    1. Packet switching (Data transmission) 2. Digital
communcations. 3. Cellular radio. I. Russell, Travis.
II. Title. III. Series.
TK5105.3.A37   1996
004.6'6—dc20

                                                    96-28768
                                                        CIP

McGraw-Hill

*A Division of The **McGraw·Hill** Companies*

ISBN 0-07-000600-8

*The sponsoring editor for this book was John Wyzalek, the editing supervisor was
Stephen M. Smith, and the production supervisor was Suzanne W. B. Rapcavage.
It was set in Century Schoolbook by Estelita F. Green of McGraw-Hill's
Professional Book Group composition unit.*

Printed and bound by R. R. Donnelley & Sons Company.

McGraw-Hill books are available at special quantity discounts to use as premi-
ums and sales promotions, or for use in corporate training programs. For more
information, please write to the Director of Special Sales, McGraw-Hill, 11 West
19th Street, New York, NY 10011. Or contact your local bookstore.

 This book is printed on recycled, acid-free paper containing a minimum of 50% recycled, de-inked fiber.

*This book is dedicated to my parents, my wife,
Marie, and our beautiful daughters, Amy
and Susan, who believe that "See Dee Pee Dee" is a
"place" that I fly to on airplanes "so Daddy can work
and buy us candy." I love you all.* J.A.

Contents

ABOUT THE AUTHORS

JOHN AGOSTA is principal of Agosta & Associates in Chicago, Illinois, a training services company for the international communications industry. A recognized authority with more than 22 years' experience in data network support, design, and management, he developed Ameritech's CDPD training systems for in-house personnel, business partners, and customers, and often leads the training sessions.

TRAVIS RUSSELL is a manager in the product development department of Tekelec. An expert in switching protocols and technologies, he is the author of *Signaling System #7*, published by McGraw-Hill.

Preface

Twenty years ago, in the mid-to-late 1970s, the state of the art in computing provided users with data transmission rates of 9600 bits per second over specially conditioned leased facilities and 1200 bits per second over direct distance dial lines. Transferring 10 million bits per second over unshielded twisted pair was unheard of. Personal computing? Six more years would pass before the first generation of PCs became available.

Revolutionary changes in the way we worked began in the mid-to-late 1980s through the use of personal, or desktop, computers. Today, most of us use a desktop machine in the office, and many of us have computers at home that are used for both business and family purposes. Our children learn their lessons at school with the aid of a computer. When someone providing us a much-needed service says, "I'm sorry, the computer is down," we wonder how we ever got along without computers. Surely, the use of personal computers can be considered one of the great revolutions of modern times.

With the proliferation of desktop computing, a second revolution was born. This was the networking revolution. Ethernet, token ring, FDDI, ISDN, SMDS, frame relay, T1, T3, SONET, and ATM, the technologies we use (and sometime take for granted) to connect individuals, workgroups, and the enterprise, are constantly evolving. Over the past ten years higher and higher transmission speeds and capacities have been achieved. The magnitude of change that has occurred in this time frame is truly amazing.

So what is the next "megatrend"? Many feel that it will not be concerned so much with speed as it will be with freedom. Wireless, mobile-capable communications systems will free us from physical boundaries and enable us to work and play where and when we need to. The cellular digital packet data (CDPD) network will have a major role in delivering the promise of the wireless revolution.

The purpose of this book is to introduce the reader to the CDPD network in a tutorial fashion. The book omits many of the finer details associated with CDPD operations and instead addresses the functionality of the network's major components and features. Included with an explanation of the functionality of each component is a detailed discussion of the protocols that are utilized to enable each component to provide the appropriate services. This book gives a nonengineering view of networking with CDPD. After reading it, you will have a solid understanding of how mobile users access, register, authenticate, and use the network. You will see how packet data and voice services are able to share the same channel space. You will understand how CDPD mobile devices learn about the network's topology, and how mobile devices can use this knowledge to perform channel and cell transfers without any involvement from the network operator.

This book is not intended to replace the various cellular and networking standards employed by CDPD. While industry standards are always the best place to look for specific details, using standards as a sole reference often leaves the reader with loose ends that need to be tied together. Here, some of those loose ends are tied together, and explanations are provided as to why various procedures are in place. If you have a need for a systems-level view of CDPD presented in a tutorial fashion, this book will have value to you. It is hoped that you not only find it informative, but pleasant reading as well. Much success to all of you!

Acknowledgments

In addition to the folks at the CDPD Forum, there are a few individuals and organizations that I would like to thank for their contributions to this book. First, thank you to the many helpful individuals at Ameritech Cellular Services who shared their insights and ideas over the past two years, most notably Scott Evans, Ron Powell, Jeff Bohlman, and Derek Podobas. Others whom I'd like to express thanks to include Jim Baichtal of AirLink Communications, for his help with analysis issues; Ray Dooley and the group at ICM; and Tim Wong of BPSI, for his assistance with M-ESs. A special note of gratitude goes to my friend and neighbor Tim Gilles of Ameritech Services, whose small acts of kindness made this book and other endeavors possible. Last, I thank Travis Russell for all of his contributions and hospitality, and a friendship that comes by only once in a long, long while.

John Agosta

1

Introduction to CDPD

Our society has changed dramatically over the last several decades. Technological advancements have affected the way we work and play. The most significant change we have witnessed in the workplace is in mobility. Middle management and executives alike are much more mobile than in previous corporate structures, as they struggle to keep up with the daily issues affecting their company.

Communication has played an important role in the mobilization of the workforce. Most of us have used pagers at one time or another, to remain reachable. However, this is not enough: There is also a need to communicate with others from any location. This need spurred the birth of the cellular telephone.

The radio telephone, predecessor to the cellular phone, had far too many limitations to meet the demands of today's workforce. With the arrival of the cellular telephone, a new industry was born. The cellular telephone industry is the fastest-growing industry in the telecommunications sector, yet the cellular telephone network is still not adequate to meet the needs of today's mobile workers.

Managers, salespeople, and executives move from one location to another, crossing the boundaries of cellular networks. Such "roaming" has placed new demands on cellular providers. The deployment of IS-41 and Signaling System #7 in the cellular network has answered the demand for roaming services. This too continues to evolve and grow. Although IS-41 resolves roaming issues, there remains a formidable constraint.

The cellular telephone is now capable of roaming most anywhere, yet the computer remains an office machine, unable to communicate without some form of attachment to the public switched telephone network (PSTN). For most people, this means using modems attached to telephone lines.

Cellular modems attempt to address this problem by allowing mobile workers to use their cellular telephones to transmit data, much like using a modem from a hotel or home. Unfortunately, the cellular network is limited to data rates of 9600 baud, which may be sufficient for electronic mail (e-mail) but terribly slow for much more. Newer cellular modems using compression can provide higher bit rates, much as they do for the circuit switched PSTN. However, this is accomplished by removing bits from the original transmission, and not by increasing the bit rate.

Mobile workers are able to get electronic mail and download files to the home office, but at a premium, because they must pay for the airtime. At 9600 baud, large files take a lot of airtime. Even smaller electronic mail files can be costly to send over the cellular network, because of the airtime.

The solution, then, is to find a technology that will support higher data rates, and that can be billed by usage rather than airtime. Even more crucial is the requirement that it be compatible with the networks already in place, that it not require costly capital to deploy, and that it support the applications already in use by much of the workforce.

The solution is cellular digital packet data (CDPD). Based on the Internet protocol (IP), CDPD provides data rates of 19.2 kilobytes per second; security through the use of encryption, compression, and a security management entity (SME) protocol; and compatibility with most corporate networks (more and more companies are connecting to the Internet for electronic mail and file transfers). The question remains how to use CDPD to improve the productivity of the mobile worker. While many major companies have signed up for CDPD service, many access providers are waiting to see how successful CDPD will be before investing in their networks. Let us look at the applications using CDPD today, and some future applications that could benefit from this new technology.

Remote Electronic Mail

Probably the most obvious application for CDPD is remote e-mail access. Mobile workers can send and receive electronic mail from any location where cellular and CDPD services are provided. In many cases, cellular providers are teaming up with Internet service providers (ISPs) to provide access to the Internet, allowing cellular subscribers to send and receive e-mail through their Internet accounts.

By connecting to the Internet, many other services become available. File transfer and remote host access are excellent tools for those on the move. Salespeople can access corporate databases to send or retrieve vital customer records, and even send sales information without having to wait until the end of the day when they return to the office. With this level of access, workers can stay in the field, where they are the most productive, instead of wasting precious time in the office performing administrative functions.

Being able to receive electronic mail while in the field allows mobile workers to keep in touch with their clients and with their office on an hourly basis. Transactions can be finalized almost in real time, providing a competitive advantage. CDPD is ideal for e-mail, because e-mail uses small packets and is "bursty" in nature, not requiring long periods of transmission.

The advantage that CDPD brings to this type of usage is the ability to leave the modem connected to the Internet all the time, rather than dialing in when one wants to retrieve e-mail. Many of today's e-mail applications provide some method of notification when an e-mail message is received, as long as the terminal is connected to the server. This is costly when using standard cellular modems, because the user is billed for airtime, but this is not the case with CDPD.

Remote File Transfer

File transfer is one of the more popular utilities of the Internet, but it is not a well-suited application for CDPD. The capability is certainly there, but file transfers tend to be lengthy. Providers charge for the number of packets sent, rather than airtime, which can make large file transfers very expensive.

If cellular providers change the pricing mechanism—say, by charging for actual airtime on file transfers—then this application might be well suited for CDPD. In the uplink (mobile to fixed station), file transfer applications can be beneficial, provided the files are small. For example, a salesperson may send an account file to the home office. Sales entries can be made by using forms on laptops, which are then transferred by way of CDPD to the home office computer. The order entry department can then enter the sales entries into the accounting system for manufacturing and accounting, or the mobile salespersons can enter the data themselves by using TELNET to connect to the accounting system.

The salesperson may wish for particular customer files to be transferred to a portable printer. For example, a customer complains that he or she did not receive an invoice. The salesperson can pull the invoice from the home office accounting system and use file transfer protocol (FTP) to transfer the file to a portable computer, from which it can be printed for the customer.

In this way a real estate agent can print information about a new home while showing a client the property. A salesperson can transfer contracts over the Internet, complete them on a laptop, and send them back to the home office. Police agencies can send artists' renderings to officers in squad cars, transfer criminal records or other reports to mobile units, even send building blueprints to fire department command posts during major fires.

While file transfer may be too costly for most applications, there are some applications where file transfers over CDPD services are viable and cost effective. As pricing structures become more competitive, this may change.

Telemetry

Utility companies are excellent candidates for CDPD services. Today, many utility companies use the cellular network to shut off appliances in homes that participate in energy conservation programs. A receiver is placed at the home, attached to the air conditioning and sometimes water heaters. When the utility experiences a peak load, it sends a signal out to all receivers to shut off the appliances for a given amount of time.

It would be just as easy to attach a transmitter to the power meter to read the meter in real time. This application has long been discussed by many utilities considering ISDN deployment, although very few have actually deployed such systems.

The advantage is obvious. With a wireless meter reader, readings can be taken anytime, even in real time. By having access to real-time data, utilities can study usage trends much more effectively, and plan in advance for sudden usage demands. Today, utility meters are read manually, by sending an individual to every home to read the meter physically. This is costly to the utility, which of course passes that cost on to the consumer. Wireless telemetry could save utilities the cost of expensive labor.

Alarm companies are also using cellular for emergency response systems. Before cellular, alarm companies depended on a landline to the business or residence to send alarm information to a remote monitoring station. This is not a dependable method, however, because telephone lines can be cut. However, a cellular transmitter can be easily hidden within a house or office, and by the time an intruder locates the cellular transmitter, it has already sent its signal.

Such systems are available today, and while they are a bit more expensive than standard alarm systems, they are more dependable for reporting to remote monitoring stations. Most of these services today use the circuit switched cellular network, but CDPD is a better choice because it offers security at low cost.

Automobile security systems can also take advantage of CDPD modems. Imagine an alarm system that could activate a modem in the vehicle, attached to a tracking system which transmits a signal over the cellular airwaves. The signal could include the vehicle identification number, license number, and owner registration information. State agencies would receive stolen car information while the car was being stolen. By using cellular monitoring systems, the signal could even be tracked to determine its location to within a few meters. Such monitoring systems exist today, and have been used by law enforcement agencies to catch drug dealers and other fugitives on numerous occasions.

Point-of-Sale Systems

Retailers often face the dilemma of trying to get telephone access to remote areas when running special sales or events. Telephone lines have become important to retailers for credit card authorization and check validation. The ability to use a wireless modem from almost any location can be a valuable service to many vendors, especially those who do not operate from storefronts. Imagine being able to pay for a pizza delivery with your credit card!

Cellular modems already exist for such applications, but again they depend on the circuit switched cellular network. CDPD offers faster access at reduced cost to the vendor. Since credit card transactions are small in file size, CDPD is a natural choice for retailers on the go.

CDPD provides mobility to go where circuits do not, which can be an advantage even in the conventional retail industry. To connect point-of-sale (POS) registers, retailers have to string lines around their establishments; and once the line has been run, the register stays put. With alternatives such as CDPD, registers no longer have a leash, and are free to roam wherever they are needed.

This applies to the transportation industry as well. Taxi companies are already using this technology to send fare information, routing instructions, and other data that would otherwise require radio transmission. Fares can be paid by credit card (or even debit card) much more easily and quickly than by manual methods. Paying by credit card has advantages: It assures that passengers are not ripped off by cabbies who overcharge and then pocket the extra cash, and that fleet operators are not similarly shorted by unscrupulous drivers.

Limousine services often rely on paper to handle credit card transactions, with no means of verifying a credit card without calling in to the office and having someone validate the credit card. With a CDPD modem in the car, the driver can validate the credit card and run the transaction through a mobile computer. This is much more efficient and provides a lot more flexibility to the limousine service. The same is true for airport shuttle services that use vans to pick up several customers at a time. The ability to perform credit card transactions from the van allows the driver the freedom to pick up any passengers at the airport and verify their credit cards, without previous arrangements having been made.

While we often think of cellular as an alternative for communicating from remote places, in many cases cellular can be an attractive wireless alternative to local area networks (LANs). CDPD offers many opportunities for using the cellular network as a wireless network in places where wiring a LAN might not be feasible.

Remote Database Access

Transmission control protocol/Internet protocol (TCP/IP) supports TELNET, a protocol used to access remote hosts. When computing from one's desktop, one usually has a LAN connection to company hosts. When on the road or at another site, the only access one may have is by dialup modem or wireless access. CDPD is already being used in a variety of applications for remote host access. Because of its widespread use for this purpose, we will discuss several applications.

American Airlines has recognized the value of using CDPD for accessing its Semi-Automatic Business Research Environment (SABRE) system. Contracted through McCaw Cellular, American Airlines is using CDPD in their terminals at several major airports for ticketing at the airlines gates. Along with remote ticketing, American can also provide gate and flight information without having dedicated terminals. This allows the airline more flexibility in which gates it uses.

American is providing customer service agents with mobile terminals equipped with CDPD modems, allowing them to access the SABRE system from any gate. Passengers looking for connecting flights or flight information can get the information they need right at the gate, instead of having to stand in line at a customer service window with hundreds of other passengers from canceled or delayed flights.

American's plans include providing kiosks in key locations within major airports that will allow travelers to access SABRE and obtain flight information themselves. Outside the terminals, CDPD will be used in the airline's fleet of service vehicles. Fueling trucks will be able to enter fueling data while they are dispensing the fuel, instead of using radios or paper records. The airline will be able to track critical fueling data in real time.

Luggage at the airport can be tracked as well, by using CDPD terminals to access the baggage handling database. Baggage handlers will use rugged handheld devices to access the system, while vehicles will be equipped with portable computers and CDPD modems.

CDPD will even hit the cruise lines, as American uses CDPD modems to ticket passengers disembarking from cruise ships (something that is done today without CDPD). The airlines are also investigating the possibility of providing onboard computers equipped with CDPD modems by which passengers could reach their office computer systems.

The airlines are not the only ones to recognize the advantage of CDPD for remote host access. Rental car agencies are also investigating the use of CDPD for their shuttles. Today, such agencies rely on radio communications to alert pickup centers that passengers are en route. With CDPD-equipped terminals on the shuttle bus, drivers can enter the customer's name, which will update a database at the pickup center and alert personnel to prepare a vehicle. Radios will no longer be necessary, and tracking customers will become much easier.

Some rental car agencies are already using terminals with cellular modems. They are presently using other cellular data networks, such as RAM Data or SMR, but as CDPD gains widespread deployment, many feel they will switch to CDPD.

Law enforcement agencies are using CDPD-equipped terminals in their vehicles to access police databases. This allows them to search for warrants and obtain driving records and license information without tying up a radio dispatcher. It also allows closer tracking of traffic calls, with transactions stored in a central computer, eliminating much of their paperwork.

An automatic vehicle location (AVL) system allows dispatchers to track police vehicles and dispatch backups without having to broadcast over the radio, providing additional security to police communications. Radio channels can now be freed for important calls where voice communication is critical.

Many more applications involve remote host access. CDPD can be a major advantage to mobile workers who need to access computers remotely, as long as the transactions are short and bursty in nature. Using CDPD to access, say, the World Wide Web (WWW) on the Internet could prove to be a costly venture, especially where graphics are involved.

As one can tell, there are many different applications where CDPD can be an efficient and cost effective alternative. Bank teller machines, package delivery companies, couriers, sales organizations, government agencies, transportation companies, and petroleum companies are all finding that CDPD fits their needs. The biggest advantage is that a user can stay connected all day long, and pay only for data sent.

CDPD Billing

As mentioned earlier, CDPD is more cost effective than the circuit switched cellular network, mainly because of the billing system. With a circuit switched network, users pay for actual airtime. This means that as long as they are connected, they are paying. There is no usage billing. This means that for short bursts of data, a user would have to pay for a minute's worth of airtime. This is true even though the transaction itself only requires a few seconds to be completed.

CDPD users pay by the kilobyte instead of airtime. This means that they can stay connected as long as they wish, and pay only for the actual data being transmitted. The disadvantage is that transmissions of large files can be costly, because one pays for actual data sent. Pricing fluctuates from one provider to the next, but most pricing falls between 4¢ and 16¢ a kilobyte, depending on contracted traffic volume. Some providers charge a fixed monthly fee of anywhere from $15 a month (for 50 kilobytes of transmission) to $50 a month (for 500 kilobytes of transmission). One kilobyte of transmission is equal to about 1000 ASCII characters.

The cost of CDPD varies depending on the organization and its usage. A recent report in *Wireless* magazine shows that field service organizations average from 200 to 800 kilobytes a month, at a cost of $38 to $74 per month. Field sales organizations use a bit more, from 300 kilobytes a month to 1250 kilobytes a month. This would cost from $50 to $110 a month. Fleet management organizations use around 100 to 500 kilobytes at $23 to $50 a month.

A simple credit card authorization costs about 3¢ using CDPD. In comparison, wireline services for a POS run to about $60 per month in line charges. Most CDPD transactions can be completed in under 5 seconds. CDPD can provide large savings in both dollars spent and time saved.

The Cellular Network

To understand better how CDPD fits into the cellular network, we must first understand how the circuit switched cellular network works. The cellular network here in the United States is predominantly analog, but this is changing rapidly as companies move toward all-digital technologies. However, even with an all-digital network, data is still sent over circuit switched networks, at the lower speeds of modems rather than the faster speeds of packet switched networks.

A number of packet switching services are being offered by cellular providers, allowing subscribers to send data over packet switching networks rather than the circuit switched cellular network. Some of these services, such as RAM Mobile Data and ARDIS, are proprietary.

CDPD is a new entry into this market, which provides a nonpropri-etary service that is compatible with many corporate networks.

In the United States, the Advanced Mobile Phone System (AMPS) has been deployed to replace the radiotelephone system. AMPS was first introduced in 1983, and was a great improvement over the radiotelephone system. Radiotelephone was limited because of the number of available frequencies, and the lack of frequency reuse.

The concept behind cellular was to use the same frequencies repeatedly, but with smaller areas of coverage (to prevent co-channel interference). By creating small areas of coverage using seven blocks of frequencies means that the same frequencies can be used again in neighboring areas, as long as the coverage areas directly next to one another are not using the same frequencies.

These areas of coverage are referred to as *cells,* and while they are usually depicted as octagonal-shaped areas, they are much more irregular. In fact, they rarely take the shape of even a perfect circle, because of the properties of radio antennas. As shown in Fig. 1-1, a block of frequencies (block A) is used to cover one cell. This block of frequencies can then be reused in a different area, because there is ample distance between the two areas of coverage (cells). The dis-tance covered by one cell can be as small as 1 mile (or smaller in some cases) or as large as 10 square miles. In major cities, where there are many subscribers, the intent is to make the cells small (microcells), so that the frequencies can be reused many times (hence more simulta-neous users). In more remote areas, the number of subscribers is usu-ally not an issue, so larger cells can be used.

Cells are often divided into sectors, with directional antennas oper-ating on different frequencies providing concentrated coverage in spe-cific directions within a cell. By deploying antennas in this fashion, coverage can be controlled closely, and transmission can be improved. When a mobile moves from one sector to another, it becomes necessary to hand off control from one antenna to another. This is facilitated by the base station controller (BSC), which resides with the antennas.

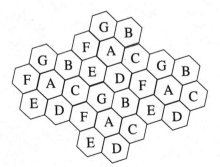

Figure 1-1 Cellular network with frequency reuse.

Each of the cellular markets is labeled as a metropolitan statistical area (MSA) or rural service area (RSA). Two frequency blocks are assigned per MSA/RSA (A and B). A forward control channel is used to "page" mobiles (indicating an incoming call) from the mobile switching center (MSC). A reverse control channel is used by the mobile to initiate a call. In the AMPS network, there are 21 "setup," or control channels within one frequency block. Each control channel is assigned to a specific cell site. To keep it basic, when a mobile phone wishes to place a call, it begins searching through the 21 control channels for the strongest signal. The phone then locks into a strong signal and sends a request for service signal. The receiving cell site uses the control channel to send the mobile phone a voice channel assignment (of which there may be thousands). The phone then begins sending voice over the voice channel. This channel remains dedicated to the mobile phone until it moves out of the cell, at which time the transmission must be handed over to another cell site (or another sector within the same cell).

If there is an incoming call, the process becomes a bit more complicated. The caller dials a telephone number which is assigned to a specific mobile switching center. The call is then routed by the telephone company to the proper MSC, which must then determine how to route the call to the mobile unit. This is accomplished through a series of database queries.

Each MSC has access to its home location register (HLR) and visitor location register (VLR), both of which provide the MSC with information regarding the location of the subscriber assigned to the mobile telephone number and service information. The MSC queries the HLR first, to determine the mobile identification number (MIN) assigned to the telephone number and whether the subscriber is registered in the network (the mobile phone has been switched on).

The HLR then identifies which MSC is currently controlling service for the subscriber. This, of course, depends on where the mobile subscriber is located. If the mobile subscriber is in the HLRs network, the HLR can query the VLR, which will identify which cell the mobile subscriber is registered in. This information is then passed to the MSC.

If it is determined that the subscriber is in another MSC, the call is routed to that MSC, which then must determine if the subscriber is still registered, and which cell is providing service to the subscriber. The MSC queries its VLR, which determines if the subscriber is registered and which cell site is currently servicing the subscriber's phone. The subscriber does not have to have the phone "off hook" for this to be determined, because the cell phone is constantly sending out a signal to the MSC for location management whenever it is powered on.

The call is then routed to the cell site, which pages the telephone by

sending out a special signal over the control channel. The subs
phone receives this signal, establishes a connection on the
voice channel, and the phone begins ringing. As one can see, lo
management can be a complex task.

A hand-off can be a tricky and complex task as well, especially
when sending data. There are two types of hand-offs, hard hand-offs
and soft hand-offs. As the mobile terminal is moving through the network, its signal may be received by more than one cell site. Each cell
site receiving the mobile's signal reports the signal strength to the
MSC. The MSC then determines which cell site is receiving the
strongest signal.

When the MSC determines that another cell site is now receiving a
stronger signal, the MSC sends out a control signal to the new cell
site, requesting a channel assignment. The new cell site then identifies an available voice channel for the call, and reserves that channel.

The MSC then sends the voice channel assignment to the mobile
over the control channel. The mobile unit then begins transmitting
over both voice channels simultaneously, so that the new cell site can
begin serving the mobile unit. The MSC sends a control signal to the
old cell site, telling the cell site to terminate service to the mobile.
The new cell site receives a hand-off message from the MSC, and
takes control of the call until the subscriber either hangs up or moves
into another cell.

This is a simplified view of how hand-offs occur in existing cellular
networks using IS-41 signaling. A few more processes must also take
place, but for this discussion we will keep it simple. CDPD uses idle
voice channels, but voice transmission always has priority in all cellular networks. For this reason, in the event a subscriber requests service from the MSC, a CDPD call in progress may be forced to surren-
der its voice channel and seek another idle voice channel. This is
referred to as *channel hopping*.

The hand-off procedure used to handle CDPD calls is different from
what we have just described. CDPD provides its hand-off procedure
(discussed in Chap. 8) when the mobile unit moves from one cell site
to another, or when the mobile is forced to seek another idle channel.
The objective is to provide consistent, uninterrupted service, regardless of whether the mobile is sending voice or CDPD data.

The New Digital Network

The cellular network is now changing over from an analog structure
to one that is all digital. There are many obvious advantages to using
a digital network, and we will examine each of these advantages.
Before we examine the technologies being used for digital networks,

however, we must first understand where digital deployment is taking place.

The cellular network is primarily digital (using digital landline facilities from the cell site through the MSC out to the public switched telephone network). However, the air interface, which goes from the mobile unit to the cellular antenna, is not currently digital. This is where new transmission techniques are being deployed to support an all-digital cellular network.

There are two methods of digital transmission over the air interface: time division multiple access (TDMA) and code division multiple access (CDMA). Both have advantages and disadvantages.

Time Division Multiple Access

TDMA uses the same frequency assignments as the AMPS network. The difference is the transmission mode. In TDMA, three subscribers can transmit on the same frequency simultaneously, just like using a twisted pair to carry multiple conversations (TDM). In the PSTN this is known as *pair gain*. A more appropriate term to use here would probably be *channel gain*. The advantages are increased revenues for each frequency as well as an increase in supported subscribers within one cell. Existing cellular telephones can operate in both the analog AMPS mode and digital TDMA mode.

TDMA is the oldest of the digital standards, developed by the Telecommunications Industry Association (TIA) and confirmed by the Cellular Telephone Industry Association (CTIA) in 1992. Deployment of TDMA networks began shortly thereafter. While TDMA offers three times the transmission capacity of older systems, the CTIA is demanding much higher channel gain of at least 10 simultaneous transmissions per channel. This requirement has not yet been met.

Call handling procedures for TDMA are somewhat different from those for CDMA. In TDMA networks, hand-off procedures are always "hard" hand-offs. This means that when a mobile moves to another sector within a cell, or into another cell, the MSC initiates a hand-off and the new cell immediately assumes control of the call. Here is how it works.

As a powered-on mobile phone (or modem) moves within the network, it sends a signal over the control channel at regular intervals. This signal is received by all antennas within its range. These antennas may be part of the same cell (sectors) or in other cells. As the signal is received, the cell site sends the control information to the MSC, along with the cell identifier and signal strength. The MSC analyzes all the received signals for a mobile unit (identified by its mobile identification number), and determines when a hand-off must take place.

If the MSC determines that a signal being received by a new cell site is stronger than the signal reported by the controlling cell site, the MSC sends a hand-off request to the new cell site. The new cell site assigns a voice channel to the mobile unit and sends the assignment information to the controlling MSC.

The MSC then sends a signal over the control channel to the old cell site, instructing it to prepare for a hand-off. The old cell site forwards the new channel assignment to the mobile, which now begins transmitting on both frequencies. The MSC then instructs the old cell site to terminate services to the mobile and release the old frequency. The new cell site is informed to begin processing the call, and the call is switched to the new cell site.

The mobile unit usually has no indication that a hand-off took place, other than a short pause in transmission. The hand-off procedures described earlier are used today in AMPS networks as well, and are facilitated through a signaling protocol, IS-41.

Code Division Multiple Access

Code division multiple access is a newer U.S. standard for digital cellular networks. Many favor CDMA over TDMA for several reasons. One major reason is increased capacity. Some users report 10 to 20 times the capacity of TDMA networks. While initial deployment of CDMA provides the same channel capacity as TDMA, extended CDMA will provide 20 to 40 calls per channel. This can be attributed to CDMA's efficient use of the frequency spectrum.

While TDMA networks place transmissions into specific channels, CDMA spreads transmissions over the entire spectrum used within the cell area. This means that transmissions are not contained in one contiguous channel, but are spread out along with many other transmissions. Special digital coding is used to allow mobile phones to differentiate among the various transmissions.

In TDMA and AMPS networks, synchronization between the mobile phone and the call site antenna is critical to maintain channelization. To achieve this synchronization, oscillators are placed in the handsets to establish clocking. These oscillators produce interference with any electrical devices within the proximity. We are all familiar with problems associated with cellular phones and aircraft navigational equipment, pacemakers, and hearing aids.

CDMA networks do not use channelized transmission, and do not need oscillators to maintain synchronization over the air interface. By eliminating the oscillators in the handsets, interference can be eliminated, making CDMA handsets safe to operate in almost any environment. This is a significant advantage over the current AMPS systems and TDMA networks, and provides new opportunities for wireless applications.

Another advantage of CDMA is voice quality. In TDMA networks, as a mobile nears the edge of a cell's boundary, quality begins to degrade. A hand-off does not occur until the MSC detects a stronger signal from another cell site and then orders a hard hand-off. In CDMA networks, the cell site can provide a "soft" hand-off. A soft hand-off is used whenever a mobile is moving from one sector of a cell to another, or as a mobile moves between two cell sites.

In a soft hand-off, it is possible for two antennas (or more) to process the mobile's signal simultaneously. Within a cell, the base station controller directs the antennas controlling the call. The antenna with the strongest signal handles the transmission until another antenna receives a stronger signal, but the MSC does not request a hard hand-off. It is possible for two antennas to switch control of the call back and forth as the signal changes. By using a soft hand-off, CDMA can provide better transmission quality as the mobile nears the boundaries of a sector or cell.

This technique is quite advantageous around the boundaries of a cell or cell sector. These areas are typically where transmission degradation is most noticeable. Both antennas covering these boundaries will experience weak signals, resulting in transmission fallout and excessive noise. When terrain fluctuates, the problem becomes more difficult to control. The soft hand-off allows both antennas to support the transmission from the mobile subscriber, eliminating fallout caused by one antenna losing signal strength.

CDMA also provides better power control. Unlike TDMA, CDMA can send control signals to mobiles, instructing them to increase transmission wattage or decrease transmission wattage, based on the received signal. Power control makes CDMA a better alternative for all wireless applications, allowing mobiles to use less power when they are close to the receiver (this is especially true when CDMA is used within buildings for wireless PBX networks). By using power more efficiently, CDMA offers longer talk times and longer battery life.

The CDMA handset is a very complex device, with high processing requirements. The handset receives all transmissions over a designated spectrum, and must be able to decode each transmission to determine if the signal should be processed or not. Each transmission is coded with a special digital code that allows the handset to correlate the many transmissions it receives. This was a deterrent to early CDMA handsets, but with the decline in computer cost and increase in computer chip processing power, this is no longer an issue.

With extended CDMA, higher bandwidth will be supported. Data rates up to 76.8 kilobytes per second will be possible with extended

CDMA, due for deployment around 1997. This will certainly make CDMA the choice technology for many PCS networks, and may provide an attractive data interface for wireless data transmission. It is uncertain if CDMA will be found to be more favorable than CDPD packet switching services.

A number of different services are provided by CDMA:

- High-quality voice transmission
- Packet data (14.4 kilobytes per second)
- Asynchronous data (14.4 kilobytes per second)
- Facsimile (G3 and G4)
- Short messaging
- Simultaneous voice and data (IS-95/IS-99)

Cost is a big factor in all cellular networks. This is especially true of the new PCS networks. Providers must invest a significant amount of money to purchase their licenses for PCS frequencies, before network deployment even begins. CDMA provides better coverage with fewer antennas than TDMA technology, which means lower deployment costs to network providers. CDMA is already a tough competitor of TDMA services, and may quickly win over the TDMA networks as well.

Group Special Mobile

The Group Special Mobile (GSM) standard is used throughout Europe today, and is very similar to U.S. TDMA networks. The principle difference between GSM and U.S. TDMA is in frequency allocation and the number of calls per channel (GSM supports 8 per channel). GSM was in use long before U.S. networks converted to digital, and looked like it might be the solution for many U.S. networks. However, there have been many concerns with TDMA usage in both GSM networks and AMPS networks.

There is much controversy over cellular transmission interference with other electrical equipment. Cellular phone usage is banned by all airlines because it can interfere with navigational equipment. There have also been concerns about interference with hearing aids, and even medical equipment. This is attributed to the fact that TDMA handsets must oscillate their transmitters to synchronize the time slots. This oscillation causes interference. In CDMA handsets, oscillation is not necessary, because time slots are not used. This means that the handsets can be used without fear of interference with other electronic equipment.

When Europe was building cellular networks, it had the luxury of building a new network without having to make it compatible with an existing infrastructure. This meant that new frequencies in the 900-megahertz range could be chosen. In the United States, network providers needed a system that could coexist with the existing AMPS network, which meant using the same frequencies as the analog network. Within the same frequency range, analog transmission can interfere with digital transmission if they are too close together.

To prevent interference between analog and digital transmissions, much of the available frequency bandwidth is used as guardbands. These act as buffers between the analog channels and the digital channels, and prevent interference between the two systems. In GSM systems, this bandwidth is used to support actual transmissions instead of acting as a guardband.

One unique feature of GSM is the handset itself. Subscribers insert a small, credit card size card into the bottom of the phone. This card provides the network with all the necessary subscriber data, including the subscriber's mobile identification number. The handset itself provides the equipment serial number. Subscribers can use any GSM handset without special activation by simply inserting their card into the phone. This feature keeps the handset simple, with very little processing required.

Summary

New demands are being placed on cellular networks by mobile users. Changes in the workplace have meant that busy managers and executives spend more time away from their desks. Laptop computers and cellular modems allow them increased mobility without loss of productivity or sacrifices in communications.

The analog network used in the United States today is changing rapidly. While the majority of the cellular network uses digital facilities, the air interface remains mostly analog. New technologies are being deployed to change the air interface to digital. Two technologies are currently being used to digitize the air interface: TDMA and CDMA.

TDMA uses time division multiplexing to allow multiple conversations to be carried over one frequency channel. TDMA requires synchronization between the mobile phone and the cell site to maintain proper channelization. This is accomplished using oscillators in the handsets. These oscillators are the source of interference induced by cellular phones.

CDMA does not use channelization, but spreads transmissions over the frequency spectrum supported by the cell site. CDMA transmissions are coded with a digital code, which must be deciphered by the CDMA handset. This requires additional processing power by the handsets, making them more complex than non-CDMA handsets.

Power control is another feature provided by CDMA. Cell sites can send control information to handsets requesting them to decrease their transmission wattage or increase their transmission wattage, based on the signal strength received by the cell site. CDMA handsets, as a result, have a longer battery life and extended talk times.

European networks use a type of TDMA, combined with a combination of protocols used throughout the rest of the network. Group Special Mobile (GSM) is similar to U.S. TDMA, but uses different frequency allocations and supports 8 transmissions per channel. GSM suffers from many of the same problems as U.S. TDMA.

CDPD resides outside the circuit switched cellular network. Data transmissions are routed to IP routers, and sent over a packet switching network apart from the voice transmissions. CDPD offers better data transmission than the circuit switched network, with higher bit rates and less interference. Billing is based on the number of packets transmitted, rather than transmission time.

It is unclear whether CDPD will be found to be more useful than newer CDMA data services, although many CDMA service providers are planning to use CDPD in their networks as well. While it is not suitable for long data transmissions, CDPD is ideal for small transmissions, and has been targeted for a variety of applications such as short messaging, e-mail, and remote database access.

2

Introduction to CDPD Architecture and Components

One of the most highly anticipated wireless data communications technologies in recent years, cellular digital packet data is much more than a new modem technology. CDPD is an open network architecture that is defined in a series of recommendations which reflect the collaborative work of numerous vendors in the hardware, software, and common carrier industries. Originally released in August 1993, revision 1.0 of the CDPD system specification outlined core protocols and methodologies required to share existing cellular airspace with packet switched data services. Revision R1.1 was released in January 1995. These recommendations clarified ambiguities and introduced enhancements to the original specifications. Collectively, these recommendations outline guidelines for deployment of a robust and secure public wireless network environment.

The network is intended to provide end users with a means to make use of existing corporate network resources where no physical connections to these resources are available. Note the use of the term "corporate network resources." Users of the CDPD network will be communicating via a wireless extension to application processes that are resident within existing network infrastructures. While the network may be used by an individual for personal use, CDPD is mainly a solution that addresses corporate network challenges as opposed to being a convenience for consumers.

In addition to being a wireless communications technology, the CDPD network has the capability of providing service to mobile users. On-demand access to network resources is made possible by overlaying packet data services on top of the existing cellular telephone infrastructure known as the AMPS (advanced mobile phone system) network. Using existing cellular infrastructures enables end users to perform communication tasks literally while on the move. Users with a need to transfer data can enjoy the freedoms currently associated with cellular telephone use.

For cellular operators, additional revenue can be generated by enabling new applications to take advantage of cellular resources that are already in place but perhaps unused at the moment. This is possible because CDPD definitions specify procedures for mobile data equipment to share radio resources with voice services. CDPD mobile devices can be redirected by the cellular operator to unused cellular channels in the event that a channel currently providing packet services experiences contention with a circuit switched service. In addition to being able to respond to explicit redirection notification, a CDPD mobile device has the ability to detect contention on its own. In this event, the mobile device "hops off" the current channel supporting packet data to another available (CDPD) channel. This channel hopping occurs without any assistance or commands from the cellular operator.

In order to speed deployment of the CDPD network, as well as allow end user applications to connect to and make use of the network with little or no modifications, the network incorporates well-understood, existing technologies wherever possible. New technologies and disciplines are introduced only when required in order to support functionality that is particular to CDPD. Some of these CDPD-specific functions include mechanisms for allowing a CDPD device to access an AMPS cellular channel, authentication processes, mobility management, and radio resource management. The interaction of these CDPD-specific and well-established processes provides end users with what appears to be a seamless, transparent connection to remote computing resources.

While the CDPD environment obviously incorporates the use of new technologies, the network's foundation is primarily supported through the use of "off-the-shelf" routing technology. With routers, the cellular digital packet data network appears to existing external communications network infrastructures as a peer network. Internally, the CDPD network provides a connectionless mode packet switching service for participating end users and applications via the IP (DOD RFC 791 Internet Protocol) and the CLNP (ISO 8473 Connectionless Network Protocol). Note that in some cases, user data is encapsulated in CLNP/IP messages for transport through the switched network fabric. User data, as presented to the CDPD environment, may also be in other forms, such as X.25 or the increasingly popular frame relay protocol. How data is presented to the network, and the subsequent encapsulation or tunneling of that data through the network, is

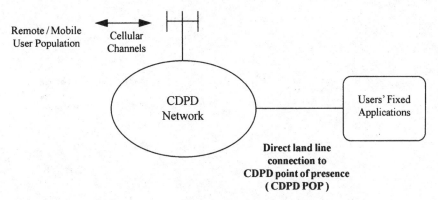

Figure 2-1 Dedicated leased facilities may provide direct connections between an external network and the CDPD network.

dependent on the protocol "stacks" user applications incorporate. The idea of user protocol stacks and subprofiles will be a topic of discussion in later chapters of this book.

From the end user's point of view (remember, "end users" can be applications or devices, such as a point-of-sale terminal), the CDPD network provides a transparent wireless access vehicle to remote (fixed) applications. In many cases, the applications that end users access are resident in external networks such as an enterprise's local area network or mainframe computer. These external networks may be directly (Fig. 2-1) or indirectly (Fig. 2-2) connected to the CDPD network. For the purposes of this book, we will consider applications involving mobile- to fixed-end-system sessions as "core" applications. While mobile-end system–to–mobile-end system communication is possible, the typical scenario will include a mobile user communicating with a fixed-end system. Some examples of mobile-end system–to–fixed-end system core applications include inventory control, order entry, e-mail, telemetry, and other query/response or transaction-oriented discipline possessing short and bursty attributes. As in any packet mode environment, file transfers and facsimile can be supported. However, core applications that generate large volumes of data are traditionally (but not always) better served in a circuit mode environment. Many factors must be considered when matching an application's need to communicate with a transport system. Some of these factors include volumes of data being sent, frequency of transmissions, distances involved, and pricing issues. For our immediate discussion, CDPD core application services will follow the traditional packet mode model of being short and bursty, with low busy-to-idle ratios or duty cycles.

In some cases, users will access applications which are resident within the CDPD carrier itself (Fig. 2-3). However, these applications

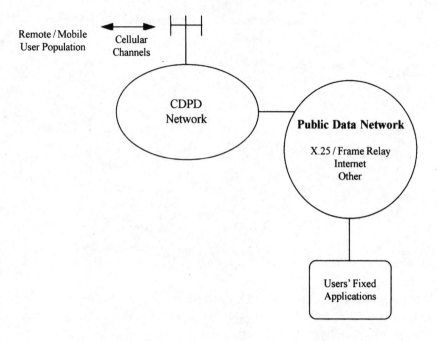

Applications can be connected to the CDPD network via third-party services providers using well-accepted technologies.

Figure 2-2 External networks may connect indirectly to the CDPD environment via third-party services.

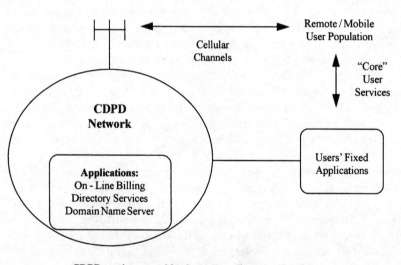

CDPD carrier may add value to "core" services by giving users access to applications that reside within the CDPD network itself.

Figure 2-3 Service providers may enable users to access carrier-resident applications.

can be considered part of a value-added service that the carrier has decided to support. Some examples of possible CDPD resident applications include e-mail services, broadcast/multicast services, directory services, or perhaps even access to on-line billing information or payment services.

Use of routing technology and open protocols makes it possible to provide what appears to be a seamless connection on behalf of the user. This is true regardless as to the location of the applications being accessed and the number of network entities collaborating with each other to provide the network service.

CDPD Network Elements

In this section we will introduce CDPD network building blocks. Each of these components and how they operate will be examined in detail in later chapters.

In its simplest form, the CDPD network can be represented by a packet switching "cloud" consisting of a set of IP or CLNP capable routers, known as *intermediate systems*. This packet cloud is the vehicle by which connected end systems can communicate (Fig. 2-4). End systems are network addressable entities that support user and/or network application services. End system entities are categorized as either *fixed-end systems* (F-ESs) or *mobile-end systems* (M-ESs). While a majority of applications will consist of sessions between F-ESs and M-ESs, the CDPD network will allow M-ES–to–M-ES communications.

Figure 2-5 depicts a network-level viewpoint of CDPD interfaces and components. We call this a network-level illustration because

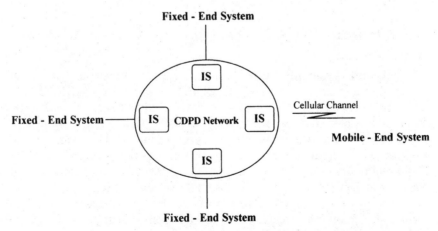

Figure 2-4 Routers provide backbone network capability for the CDPD network.

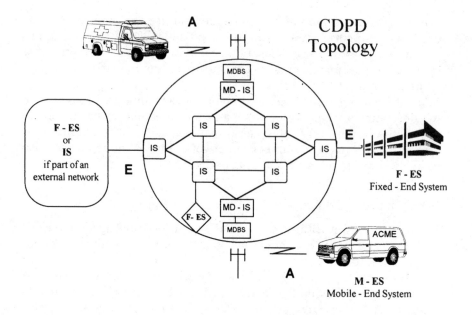

A network level view of the CDPD network.

Figure 2-5   CDPD addressable entities.

each of the components indicated are addressable entities. Because these components can be addressed, data packets can be directed to each of them, and perhaps of greater significance, they can be managed remotely. A description of each of these components and interfaces follows.

The Air Interface

The air (A) interface exists between a mobile database station (MDBS) and the mobile user population and consists of the set of radio channels utilized to support CDPD services. The MDBS is the CDPD radio control system which can be found at each cell site. Packets are delivered to a cell site via terrestrial landlines, microwave transport systems, or packet switched network services. Once they are at the cell sites, data packets are relayed onto a cellular radio channel toward the mobile user population. The radio segment consists of forward and reverse channel streams which together provide a full-duplex transmission path over the air. The access method used over the air interface is called digital sense multiple access with collision detection (DSMA/CD).

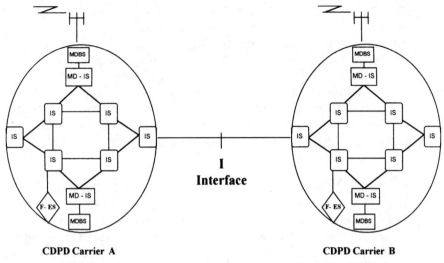

CDPD Carrier A   **CDPD Carrier B**

Figure 2-6   The I interface serves as a gateway between participating CDPD carriers.

The Interservice Provider Interface

As shown in Fig. 2-6, the interservice provider (I) interface provides gateway services between cooperating CDPD carriers. These interfaces facilitate interdomain routing of user data packets as well as network administration information such as traffic counts and other billing or authentication information. The existence of the I interface should be transparent to user applications. Therefore, no prior knowledge should be required for an application to be able to communicate over this interface, facilitating seamless interdomain routing.

The External Interface

The external (E) interface provides connection between a CDPD network and a non-CDPD network. Perhaps the most obvious example of an E interface is the connection between the CDPD network itself and end-user application systems which provide core services to mobile users. Other examples of E interfaces are connections to third-party networks such as CompuServe, America On-Line, Dow Jones News, Internet service providers, and others.

Fixed-End Systems

A fixed-end system is a network-addressable system that is connected to the CDPD service provider's network via terrestrial means. The F-ES may be external to the CDPD network or resident within the network. An example of an external F-ES is an application server which is being used for inventory control purposes, such as would be found within an airline reservation system. The external F-ES is administered and supported by the CDPD subscriber, which in most cases is an organization, not an individual. An example of an F-ES internal to the CDPD network is a directory services database. By definition, F-ESs are stationary devices. F-ESs are not required to have any knowledge of mobility when communicating with an M-ES. Other CDPD network elements should work together to ensure that all phenomena associated with mobility are hidden from the F-ES.

Mobile-End Systems

A mobile-end system is a mobile-capable network interface for users and applications. The M-ES is essentially a CDPD "modem" which maintains continuity of service when operating over the AMPS cellular radio system. While the M-ES may be stationary, the M-ES supports all the functions necessary to perform reliably in the mobile environment. Some of these functions include initial radiofrequency channel acquisition, network registration, channel hopping, and other radio resource management tasks. Because the M-ES can be considered a CDPD "modem," it should provide a transparent interface to higher-layer applications which are attempting to make use of the CDPD network.

Intermediate Systems

Intermediate systems (ISs) form the backbone of the CDPD network. They are responsible for the relaying of user data, network administration, and "reachability" information through CDPD carrier domains. ISs are off-the-shelf routers which support the IP/CLNP network-level relaying protocols. ISs also support international and Internet standard routing protocols such as Open Shortest Path First (OSPF), Border Gateway Version 4 (BGP-4), and Intermediate System to Intermediate System (IS-IS), to name a few. In some instances, an IS may support X.25, frame relay, and other communications disciplines in order to support user end systems operating in multiprotocol environments.

Mobile Database Stations

Mobile database stations (MDBSs) form the physical interface between the baseband and radiofrequency (RF) environments. Each cell site which is provisioned for CDPD service is equipped with an MDBS. The MDBS functions as the cell's CDPD radio channel controller and is responsible for the management of cellular RF resources. Data packets arriving at a cell site from the mobile telephone switching office (MTSO) are relayed onto a cellular CDPD forward channel stream for transmission toward the mobile user population. Data packets arriving from a M-ES are demodulated off the cellular CDPD reverse channel and relayed toward the MTSO via a noncellular transport system. The combination of a forward and a reverse channel stream is referred to as a CDPD channel pair. A CDPD channel pair provides remote users access to a two-way simultaneous (full-duplex) wireless transmission path.

The transmission path and facility connecting the MTSO with a cell site does not need to be separate from the transmission path currently being used by voice or circuit switched services. CDPD traffic may share the transmission facility with voice traffic via utilization of D4/ESF DS1 multiplexes, microwave, or other technologies. Figure 2-7 illustrates how CDPD data is overlaid on top of existing voice services.

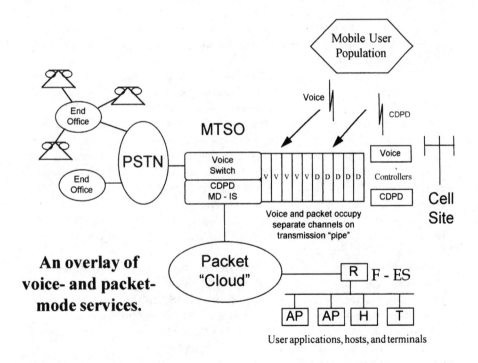

Physical bandwidth may be shared, but voice and CDPD services are supported by two separate networks.

Figure 2-7 The CDPD and voice network relationship.

Examples of RF management responsibilities carried out by the MDBS include adjusting M-ES transmit power levels, advising M-ESs to switch to a different RF channel, and monitoring for contention of an RF channel by other (non-CDPD) services.

Mobile Data Intermediate System

A mobile data intermediate system (MD-IS) is the logical interface between the mobile (air) and the fixed (backbone) environments. For the purpose of our discussion, we shall define these two environments.

Referring to Fig. 2-7, we see that the MD-IS is essentially a fence sitter. On one hand, the MD-IS converses with ISs, which are simply off-the-shelf, commercially available router products. This is the backbone environment. The protocols obeyed on the backbone side of the fence conform to well-known international standards. Routing occurs according to rules that are very well accepted, namely, the (DOD RFC 791) IP and the international standard ISO 8473 Connectionless Network Protocol. The mechanisms employed between the commercially available backbone ISs and the MD-ISs to identify "best path" information through the backbone are also well known. Examples of open processes that are used to exchange reachability information and determine best routes between ISs are the OSPF (Open Shortest Path First) protocol and BGP-4 (Border Gateway Protocol Version 4).

On the other side of the fence, toward the air interface, the MD-IS has conversations with logical entities and physical devices that are unique to CDPD. These devices are the MDBSs and, of course, the M-ESs that are currently being served within the MD-IS's "subarea routing domain." A level 1 subarea routing domain is defined as the set of cells which are under the authority of a given MD-IS. This level 1 subarea routing domain can either represent all of a carrier's cellular geographic serving area (CGSA) or just one of many such areas within a CGSA (Fig. 2-8).

In addition to some of the well-understood IP routing processes that are used over the air interface, the MD-IS is concerned with CDPD specific activities. Some of these activities include the authentication, connection management, and tracking of remote mobile users.

The MD-IS is the only router or network element that has knowledge of mobility, and because of this characteristic it can be considered the heart of the CDPD network. Resident within an MD-IS are two functions: the mobile home function (MHF) and the mobile serving function (MSF).

Each individual M-ES is logically associated with a given MD-IS acting as the mobile system's "home." All datagrams (a term used to describe data packets using connectionless network protocols such as

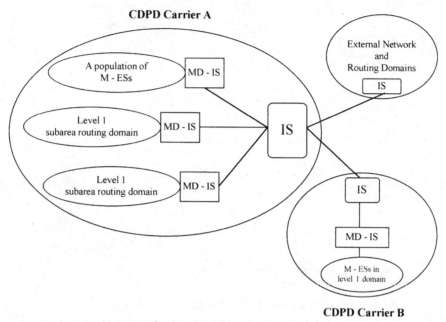

CDPD Carrier A

CDPD Carrier B

Figure 2-8   CDPD routing domains.

IP/CLNP) destined for a mobile-end system are routed to that mobile's MHF, which is resident within the mobile's home MD-IS. The MHF maintains a table of the locations (MD-IS/MSF) from which each of its home clients is being serviced at the moment. In the forward direction, the MHF then forwards datagrams to the appropriate MD-IS serving a target client. This forwarding function is facilitated by simply encapsulating the original datagram, which is addressed to the target M-ES, inside a second datagram which is addressed to the MD-IS currently providing the mobile service. This approach is analogous to the U.S. Postal Service forwarding mail to a recipient's new address. As Fig. 2-9 illustrates, one should also note that a MHF may or may not be resident within the same physical MD-IS that is currently serving the mobile. Because the CDPD network is comprised of a set of subarea routing domains from more than one service provider, and the MHF acts as a pivot point for packets traveling in the forward direction, an interesting scenario is presented.

Assume that a mobile user's home is in New York City. The F-ES that contains the inventory database that is normally accessed is located in Chicago. While she is on a business trip to Chicago to attend a sales meeting, the mobile user checks the database for some information via CDPD. While packets traveling in the reverse direction (toward the F-ES) take a direct route within the city of Chicago (Fig. 2-10), packets traveling in the forward direction (toward the

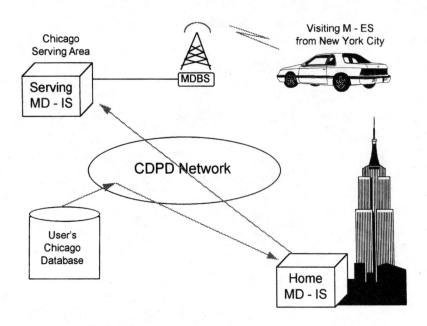

An example of data flow in the F - ES to M - ES direction.

Figure 2-9 Forward data flow.

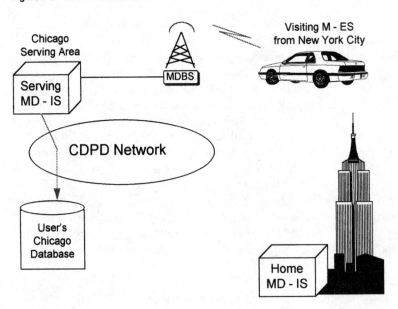

An example of data flow in the M - ES to F - ES direction.

Figure 2-10 Reverse data flow.

M-ES) must first be relayed via the MHF in New York. From the MHF in New York, packets can be encapsulated and forwarded to the mobile's serving MD-IS in Illinois (Fig. 2-9). The additional router hops, encapsulation overhead, table lookups, and other processing in the forward direction may collectively affect the application (or user's) perceived throughput and/or response times.

When "roaming" in this manner, pricing and other business agreements must be established between the participating carriers. Regulatory issues may also apply. Does the user have a choice of interexchange carriers to choose from? Does an interexchange carrier even need to be part of this scenario? Are there roaming charges? Will the user receive one bill reflecting charges for a "session," or many bills? Discussion of these topics will follow in later chapters.

Summary of Key Points

- The CDPD network is an internetwork of participating autonomous network service providers. The switched fabric of the CDPD network utilizes connectionless mode packet delivery, which can be provisioned in many cases by readily available, off-the-shelf router products. Essentially, the network is a mobile-capable IP internet.

- New technologies are introduced to the CDPD environment only when required. Generally speaking, these new technologies have to do with managing the cellular environment and mobility aspects of the service.

- The CDPD network is primarily a tool that will provide communications solutions for business needs. However, any organization, including individuals, may subscribe to CDPD services if operating costs can be justified.

- CDPD network providers may at their discretion provide enhanced or value-added services that are not resident within user end systems. Instead, these application services may be resident with the CDPD providers' network itself, or in third-party networks.

- Authentication, redirection, and forwarding of data packets to mobile users is accomplished via a mobile home function that serves as an anchor or pivot point for mobile-end systems. Because all traffic in the forward direction must transit the MHF, revenue sharing and equal-access issues must be resolved as the CDPD network evolves and matures over time. In the short term, simple dedicated lines interconnect CDPD carriers, which is perhaps a simpler and more attractive internetwork solution.

3

Layering, Profiles, and Subprofiles

Layering

The set of communications protocols and processes used within a CDPD environment will vary from reference point to reference point (see Chap. 2). In addition, applications within the network itself as well as within the users' domain will determine the "look" of a system as far as the collection of communications processes is concerned. Two terms that are used to describe the set of communications processes implemented at a network reference point are *profile* and *subprofile*.

Before examining CDPD-specific layering scenarios, we shall review basic layering principles and terminology. Figure 3-1 is a representation of the Open Systems Interconnect (OSI) reference model. While many teleprocessing systems do not follow this model exactly, the model is the basis for most systems now being deployed in the world today. An analogy is offered to illustrate this point.

Consider an audio stereo system. Certain core functions must be designed into every stereo system: signal acquisition (perhaps from disk, tape, or radio frequencies), decoding and demodulating, amplification, and air movement. Some systems incorporate many of these functions into a single component. The stereo receiver which also functions as a power amplifier is an example. However, other systems are built in such a manner that each individual function is accomplished by a separate component. A tuner mated with an external power amplifier, which in turn is mated with a 20-band graphic equalizer, is an example of such a scenario.

Similarly, Fig. 3-1 shows a system which incorporates separate sets of software to perform individual tasks. Figure 3-2 shows a system

Layer 7 Application	System - independent processes which interface between the user and operating system. Examples of services include message handling and file transfer.
Layer 6 Presentation	Description of data syntax and format. Identifies data attributes as binary, boolean, IA5 characters, etc.
Layer 5 Session	Activity management functions allowing for the control of data exchanges between end systems and sync pointing of user data.
Layer 4 Transport	End system to end system acknowledgment, flow control, and other "housekeeping" activities performed on data transfers.
Layer 3 Network	End system to end system addressing and routing mechanisms. Network - level switching of user data will occur based upon network layer protocol definitions.
Layer 2 Data Link Control	Node to node or "hop to hop" addressing and control of data transfers between end systems as well as intermediate systems.
Layer 1 Physical	Electrical and procedural processes that enable the correct interpretation of individual bit status.

Figure 3-1 The OSI model.

Modified "stack"

"Bundled" package providing application, presentation, session, and transport level services.

Layer 3
Network

Layer 2
Data Link Control

Layer 1
Physical

Figure 3-2 Multiple-layer functionality within a single software process.

that utilizes bundled software which provides internally all the services normally provided by individual OSI processes.

Primitives, Data Units, and Service Access Points

How do layers within a system communicate with each other? A well-defined set of *system calls* is implemented to invoke services between vertical layer boundaries. These service calls are referred to as *primitives*. As Fig. 3-3 implies, there are four generic primitive types: requests, confirmations, indicates, and replies. At the origination point of a session, the *request* primitive is used by a higher layer (N) invoking the services of an adjacent layer directly below it $(N - 1)$. Confirmations are issued back up the stack to the originating layer to inform that the requested services are being set up, completed, or oth-

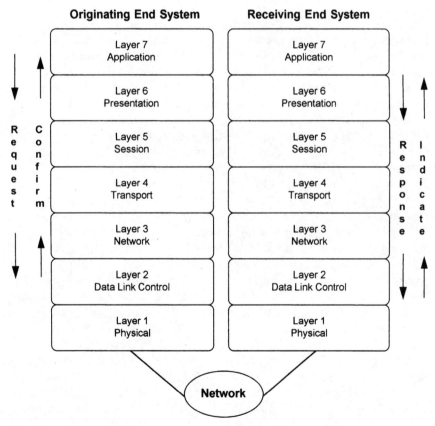

Figure 3-3 OSI primitives.

erwise processed. At the destination point of a session, *indicates* are issued from a lower layer (Y) to the layer immediately above it ($Y + 1$) in order to invoke that higher layer's services. Upon receiving an indicate from a lower layer, the $Y + 1$ layer issues a *response* to the request for services. Primitives can be considered system calls in that these messages exist solely between layer boundaries within an end system. They are not transmitted across serial communication links between end systems.

The actual information which is acted upon (processed) by a layer is referred to as a service data unit (SDU). When a layer has completed its processing tasks, the original SDU is passed downward to an adjacent lower layer with an accompanying header appended to the original SDU. The header is sometimes referred to as *protocol control information* (PCI). An SDU which is processed and handed down to an adjacent lower layer with a new header is referred to as a *protocol data unit* (PDU). Figure 3-4 illustrates these relationships. Essentially, a PDU leaving layer N becomes the SDU to be processed at layer $N - 1$.

CDPD, like any other multiprotocol-capable system, requires a mechanism that can be used to direct PDUs to the correct process for treatment. For example, data that is generated by a file transfer application should not be directed to an e-mail system at the receiving end. If a machine is multiprotocol capable, an IP datagram should be processed by a "peer" IP entity at the receiving end, not a CLNP, X.25, or any other process. The specific tool that is used to direct protocol data to the appropriate peer entity is called a *service access point* (SAP). The SAP is an address that indicates what specific entity is responsible for PDU processing, and provides a means to "point" the PDU toward the appropriate peer entities at each layer within a system. Other terms used in the industry for mechanisms that provide this functionality are *port, socket,* and *type code.* Figure 3-5 shows the "native" TCP/IP profile and where these protocol identifiers are located. A SAP (or equivalent) actually is passed through the communications channel between end systems, and is found with the PCI information of a data PDU. Within the CDPD network, datagrams must be directed to the appropriate function, be it a data relaying, radio resource management, security, or registration process. Accordingly, service access points need to be defined for each of these CDPD specific processes.

At this point, we should feel comfortable with basic data flow within an end system. Remember that primitives are used to invoke and manage service requests, while SAPs are used to direct data within a communicating system. But what are the functions of each layer within an end system? The following digest highlights the responsibilities of each of the OSI processes.

User - generated traffic = HELLO

Data to send	HELLO (SDU)	**Application**
After processing	7PCI, HELLO (PDU)	**L7 PDU**
Data to send	7PCI, HELLO (SDU)	**Presentation**
After processing	6PCI, 7PCI, HELLO (PDU)	**L6 PDU**
Data to send	6PCI, 7PCI, HELLO (SDU)	**Session**
After processing	5PCI, 6PCI, 7PCI, HELLO (PDU)	**L5 PDU**
Data to send	5PCI, 6PCI, 7PCI, HELLO (SDU)	**Transport**
After processing	4PCI, 5CI, 6PCI, 7PCI, HELLO (PDU)	**L4 Segment**
Data to send	4PCI, 5CI, 6PCI, 7PCI, HELLO (SDU)	**Network**
After processing	3PCI, 4PCI, 5PCI, 6PCI, 7PCI, HELLO (PDU)	**L3 Packet**
Data to send	3PCI, 4PCI, 5PCI, 6PCI, 7PCI, HELLO (SDU)	**Data Link Control**
After processing	2PCI, 3PCI, 4PCI, 5PCI, 6PCI, 7PCI, HELLO, 2PCI (PDU)	**L2 Frame**
Data to send	2PCI, 3PCI, 4PCI, 5PCI, 6PCI, 7PCI, HELLO, 2PCI (SDU)	**Physical**
After processing	Serial bit stream	**Binary bits**

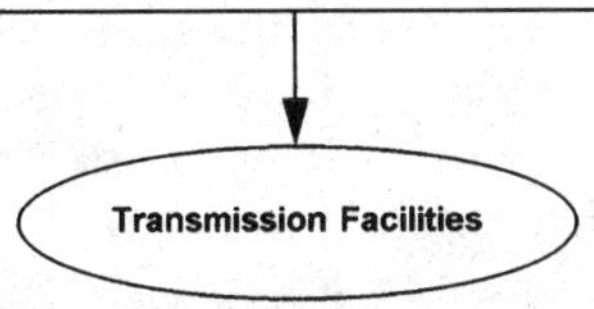

Transmission Facilities

Each layer adds more information to be transmitted, increasing the total
message length. Achieving an efficient overhead-to-data ratio is important.
Not doing so may adversely affect users' perceived performance.

Overhead encountered in a "layered" system

Figure 3-4   Layering overhead: PDU and SDU relationships.

The Physical Layer

Layer 1, the physical layer, consists of the set of mechanical, procedural, and electrical rules of operation associated with a machine's
input/output ports, and the mechanisms used to identify individual
bit status. To use a vintage example, consider a V.35 interface. Layer
1 attributes include the definitions of each pin on the connector, the

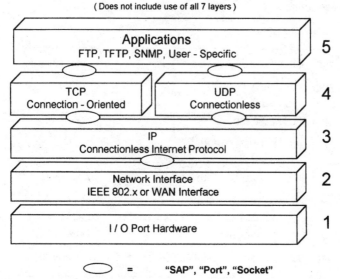

Figure 3-5   A TCP/IP end system.

"handshake" protocols such as RTS/CTS that are implemented, voltage levels, and the use of a "balanced" interface. Individual bit status identification is accomplished by comparing data leads on the connector with synchronous clock leads on the same connector.

As newer and more complex technologies were introduced, the definitions of what was considered a layer 1 attribute became more complex. The high-speed asynchronous transfer mode (ATM) and switched multimegabit data service (SMDS) provide examples. Each of these systems switches fixed-length "cells" of 53 bytes at extremely high speeds. In broadband networks such as these, there are additional processes that can be best described as "layer 1.5," lying someplace between layers 1 and 2 in the OSI model. Adaptation of user data into the cells itself, and "convergence" sublayers that map cells onto a physical transport system such as DS1, DS3, or SONET, are included in the area between layers 1 and 2. In some circles, there is debate as to where layer 1 responsibility begins (see Fig. 3-6).

When considering vintage and modern technologies, the physical layer is best described as the set of mechanisms used to assist in the correct identification of individual bit status. These mechanisms may include not only raw electrical characteristics, but also a series of framing mechanisms to enable transport facility synchronization as well as "payload" delineation.

Traditional view of physical layer

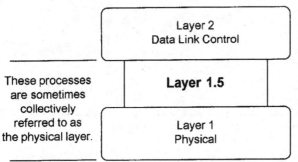

Layer 2
Data Link Control

Layer 1
Physical

Direct conversion:
Layer 2 PDUs into
serial - bit stream.

A more complicated view of the physical layer

Layer 2
Data Link Control

Layer 1.5

Layer 1
Physical

These processes
are sometimes
collectively
referred to as
the physical layer.

Intermediate conversion:
Various adaptation and convergence
functions map layer 2 PDUs onto a
physical transport facility and provide
payload delineation.

Figure 3-6 Newer technologies have changed physical layer definitions.

The Data Link Control Layer

Layer 2, the data link control layer, is concerned with node-to-node, or hop-to-hop communications. Generically, data link control (DLC) operation can be categorized as either unnumbered information transfer or sequenced information transfer. Many systems use both categories, as does CDPD. Unnumbered information transfer simply identifies the destination endpoints of the transfer, while sequenced information transfer also checks to make sure that each information frame (L2 PDU) being sent over a communications link is accounted for. By applying an inventory number (send sequence number) to each information frame that is transmitted, it is possible to detect and correct errors in transmission over a communications channel by re-sending frames that are missing or received out of sequence.

In classes that I teach, I like to use the air travel system as an analogous example. Consider taking a flight from Memphis, Tennessee, to Kalispell, Montana, as indicated in Fig. 3-7. The itinerary may begin with a flight segment between Memphis and Chicago, Illinois.

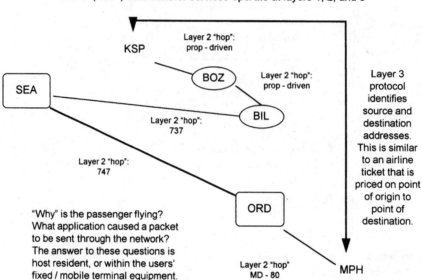

Figure 3-7   Data link and network-level functions as analogous to air travel.

Another flight segment connects Chicago to Seattle, Washington. Additional flight segments eventually get to the destination, Kalispell. Each of these flight segments can be considered a layer 2 hop. The aircraft and flight crew is concerned only with safe departure and safe arrival at each airport. They are not aware whether this is the final destination or just a connection point.

In a network environment, DLC mechanisms typically are concerned with the endpoint addressing associated with each hop, and with detecting bit errors across a hop. Usually, another layer is responsible for routing information through each intermediate station until the information reaches its ultimate destination.

The Network Layer

Layer 3 of the OSI model, the network layer, is the set of disciplines used to identify the source and destination addresses of communicating end systems as well as the set of rules that govern the communications between end systems. In our air travel analogy, a passenger (packet) is originating in Memphis (source layer 3 network address) and is terminating in Kalispell (destination layer 3 network address). The fare and other rules governing MPH-KSP travel are part of network level considerations. Common network level protocols are the RFC791 IP, the ISO 8473 CLNP, and the international standard X.25 packet level protocol. Routers forward packets based on network-level protocol elements of procedures.

The Transport Layer

Layer 4, the transport layer, is concerned with end system–to–end system reliability and accountability at the perimeter of the network environment. In other words, transport functions are *host resident,* and no knowledge of end-system transport attributes is required by the network environment in order to route data from source to destination.

What occurs during telephone conversations can be used as a suitable analogy for transport functions. The network (telephone company) does not need to know what language a telephone conversation is being held in. The communicating end systems (people) periodically acknowledge each other's dialog with "affirmatives," "say again pleases?", or perhaps just a series of snorts, grunts, or other audible sounds. The net result of all this is that regardless of whether a conversation is being held in English or Farsi, the two parties are aware that they are talking to each other, and not to a telephone dangling in the air.

Transport-level functions may incorporate segmentation, sequence numbering, acknowledgment, and flow control procedures between communicating end systems. The ISO 8073 TP Class 4 (TP4) and the Internet's RFC792 TCP are examples of such processes.

The Session Layer

Layer 5, the session layer, consists of host-resident processes that govern the interaction of communicating end systems. A good term for layer 5 responsibility is *activity management.* An *activity* can be described as an engagement between two application processes, such as a file transfer system. Within the activity, multiple users can be sending and receiving various files totally independently of each other. Each user's "connection" is referred to as a *session.* An example of activity management is layer 5 processes giving permission for one user connection to send while other sessions are temporarily on hold. This form of flow control may be used when multiple parties contend for the same application resources.

Another feature of activity management is the concept of *sync pointing.* During some sessions, especially large batch file transfers, it is desirable to append sync points among application data transfers periodically, to account for the volumes of data successfully received thus far in the session. In the event of a failure of the network fabric, this layer may be able to determine the last mutually acknowledged sync point associated with a session that has not been formally closed. Upon reestablishment of a connection, the session layer will be able to recover and continue on from the point of failure. Human beings incorporate layer 5 processes when they are disconnected during a telephone call. Upon reestablishing the telephone connection, we typically say to each other "so where were we?" and pick up from the point of interruption, as opposed to starting the conversation all over again. Elements of procedure of the OSI session layer can be found in ISO 8326 and 8327.

The Presentation Layer

Host-resident processes that are used to define the context and syntax of data transmissions comprise the presentation layer, layer 6 in the OSI model. A very simple example of a layer 6 process in action is a scenario in which a machine receives a bit sequence of 01000011. It is impossible to interpret this string without being previously informed of its context, or how it should be processed. 01000011 may be the IA5 character "C." 01000011 may also be the binary number for the decimal value "67." 01000011 may be a binary-coded decimal (BCD) representation of the digits "4" and "3," or 01000011 may be part of a Boolean expression, and this string should be exclusive ored (X ORed) with another string that has previously been received.

Obviously, without identifying the rules by which to consider the data, the data itself is meaningless. The ASN.1 (Abstract Syntax Notation.1) protocol is an available set of rules that may be applied to data communications and accomplishes this task.

The Application Layer

Layer 7, the application layer, comprises standard host-resident processes that enable a common set of tools to be used for applications such as file transfer, electronic messaging, and virtual terminal emulation, to name a few. Examples of application layer processes include ISO's FTAM and DOD's FTP for file transfer operations, as well as the ISO X.400 protocol and the DOD SMTP for electronic messaging.

Note that while ISO and CCITT/ITU consider application level processes to layer 7 entities, DOD/Internet application processes are considered layer 5 entities, because the DOD's TCP/IP profile does include the use of session and presentation services in its model (see Fig. 3-5).

Behavioral Characteristics

From layer 2 up toward layer 7 in any stack, the characteristics of each protocol process provide either connection-oriented services (COSs) or connectionless services (CLSs). Whether a protocol can be considered a connection-oriented or a connectionless protocol depends on two factors: how the service is provisioned, and how data transfers take place once a path is established.

First, let's talk about the "provisioning" of a service. If a service, or path through the network environment, must be defined at some point (perhaps at subscription time) by a human being, and configured into a network device in some manual manner, then this service is a connection-oriented service. Frame relay and X.25 Permanent Virtual Circuits (PVC) are examples of this type of provisioning. All information that needs to be sent in a virtual circuit environment (switched or permanent) travels along a fixed, predefined path.

Also, with respect to provisioning a service, if formal procedures are used to establish a connection, maintain a connection, and release a connection, then that protocol can also be considered a connection-oriented protocol. Examples of this type of operation are X.25 Switched Virtual Circuits (SVC) and both the ISO 8073 TP Class 4 and the DOD Internet TCP.

Protocols that have the ability to establish a path through a network environment without formal connect procedures on a dynamic basis, and can choose the path that is (hopefully) the most cost-effective and efficient route at the moment, are considered connectionless protocols. The "packets" that are passed in these environments are often referred to as *datagrams*. Two examples of connectionless datagram protocols are the ISO 8473 CLNP and the DOD/Internet RFC791 IP.

Datagram protocols require the aide of "partner" protocols that have the ability to discover best routes through the switched fabric, and exchange reachability information among the various switching elements. These partner protocols are referred to as *routing protocols*. Examples of routing protocols are the Routing Information Protocol (RIP), Open Shortest Path First (OSPF) protocol, Border Gateway Protocol (BGP), Intermediate System to Intermediate System (IS-IS) protocol, End System to Intermediate System (ES-IS) protocol, and Intermediate System to Intermediate System Inter-Domain Routing Protocol (IDRP), to name a few. Routing protocols typically use a set of metrics which may consist of hop counts, load factors, bit rate, and transmission facility quality characteristics when making "best route" decisions.

Next, we discuss the information transfer attributes of communications protocols. When considering how protocols "behave" when actually transferring information, connection-oriented and connectionless services vary in the ability to keep an inventory of how much data was sent, whether or not the data was received in the correct order, and the ability to perform error correction by retransmitting messages that are deemed to be missing, out of sequence, or are otherwise found to be in error. We will refer collectively to these activities as *housekeeping*. If a protocol is not concerned with such housekeeping activities, the protocol "behaves" in a connectionless manner. It is possible to have a communications protocol, frame relay being an example, that is by definition a connection-oriented protocol (virtual circuit operation) but that possesses information transfer attributes (no housekeeping) like a connectionless protocol.

The network-level CLNP and IP are two "true" connectionless protocols because not only can a path vary on a dynamic basis (datagrams), there are no housekeeping activities during information transfer associated with these disciplines.

The X.25 protocol and the TCP provide "true" connection-oriented services in that they not only invoke formal call establishment and release procedures, but they perform housekeeping, with activities such as sequencing, acknowledgments, and retransmission attempts when required.

An obvious design goal in any communications environment is to make the service as reliable as possible. A process must be incorporated into any system profile that works hard to manage the transfer of information and perform tasks such as sequencing and retransmission in order to provide such reliability. A designer typically incorporates the use of a connection-oriented discipline at one or more layers within the system profile. However, the designer must not duplicate housekeeping activities across so many layers that the ratio of time transferring information to time managing information transfer becomes inefficient.

For pure broadcast applications such as ticker tape, news, and weather information distribution, the designer has no reverse path which can be utilized for retransmission requests. In such cases, the designer must incorporate forward error correction (FEC) mechanisms which allow the receiving end system not only to detect the presence of bit errors, but also to identify what bits are in error and correct them itself. FEC is sometimes supplemented with additional procedures such as transmitting information multiple times, performing FEC calculations on each of the received messages, and using a "majority rule" criterion to discard "bad" messages and process "good" messages.

CDPD employs both connection-oriented and connectionless disciplines at various points in the network. Where transmission facilities are expected to be reliable, such as between ISs, connectionless data-gram capabilities are used. Where bit errors in transmission are expected, such as over the air interface, connection-oriented disciplines with full housekeeping capabilities are utilized. Inefficiencies created by the introduction of airlink housekeeping activities and CDPD's multiple layer subprofiles are offset by FEC calculations and data compression.

Applying Layering Concepts to the CDPD Environment

Recall that the set of disciplines implemented at each layer within a teleprocessing system defines a *profile*. Vertically within a stack, CDPD system profiles can be dissected into three *subprofiles:* an *application subprofile,* a *lower layer subprofile,* and a *subnetwork sub-profile.* An application subprofile is a set of disciplines found above layer 4 in the OSI model that consists of carrier support, management, and billing processes as well as mobile user applications. A lower layer subprofile is a collection of transport and network layer processes incorporated within the CDPD network elements. Subnetwork subprofiles reflect the nature of the connecting mediums that are implemented between communicating network level entities.

Figure 3-8 illustrates a profile set that may be used internally within a CDPD network to provide support and management services critical to the operation of the network itself. Notice the use of a connection-oriented transport service to provide reliability to the transfer of carrier essential information such as billing information, internal messaging, and network management information. All of this type of information will be routed through the network itself via the ISO 8473 CLNP. The CDPD network provider is free to choose from many available subnetwork technologies to relay this information across a given hop between network elements.

Figure 3-9 illustrates that users' network management information, in the form of the Simple Network Management Protocol Version II (SNMPv2), utilizes a connectionless transport mechanism, the Internet User Datagram Protocol (UDP) as defined in Internet standard RFC768. Routing of this information through the network occurs using the DOD RFC791 IP. Again, between individual network elements, the CDPD provider may utilize many off-the-shelf subnetwork technologies for hop-to-hop control. It should be noted that network management information sent to and from mobile-end-systems traffic consumes bandwidth on the cellular airlinks, which is considered to be a precious resource. Accordingly, inclusion of SNMP support by an M-ES is considered to be an option within the system specifications.

Carrier Support Service Subprofiles

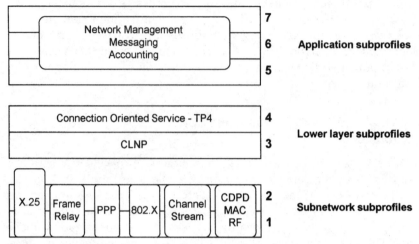

Figure 3-8   A communication profile for CDPD network support services.

User Support Service Subprofiles

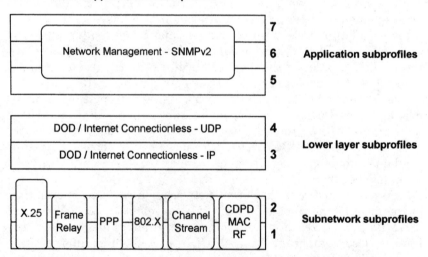

Figure 3-9   A user network management profile.

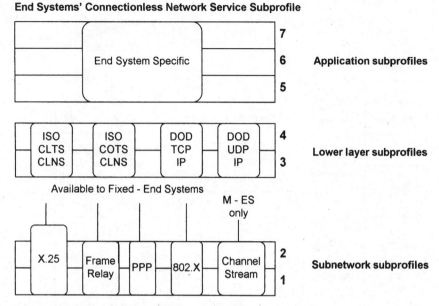

End Systems' Connectionless Network Service Subprofile

Figure 3-10   User application data transfer profile.

Figure 3-10 illustrates the M-ES specific subprofile used for user application data transfer. Note that the higher layers are not defined, as it is here where the user's applications will reside. The M-ES equipment always utilizes either the ISO or DOD connectionless network service for routing purposes. However, the user may select either connection-oriented or connectionless transport services, depending on the degree of reliability that is required for data transfer. Because the M-ESs operate over the A interface, only CDPD channel streams can be used as the subnetworking technology. Fixed-end systems, as Fig. 3-10 implies, may use any subnetwork technology to connect from the users' office to the CDPD carrier's point of presence.

The CDPD Air Interface Profile

Figure 3-11 provides a picture of the various processes used between the MD-IS and M-ESs, including the processes resident at the CDPD cell sites. The protocol layers denoted with arrows are the mechanisms that govern the air interface.

As stated in Chap. 2, the physical connections between an MD-IS and a cell site under its control depend on the topology and technologies a CDPD service provider decides to implement. Protocol layers

Airlink Protocol Profile

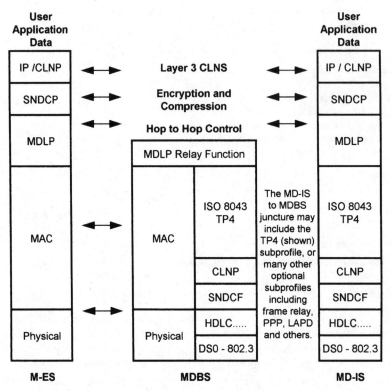

Figure 3-11   The mobile user-to-network interface.

that a carrier may decide to implement between a cell site and its controlling MD-IS are not marked with arrows on this graphic. Of particular interest is the possible use of a TP4/CLNP/SNDCF subprofile.

Recall that the TP Class 4 is connection oriented, and includes a formal call connect procedure. When TP4 issues a connection request, it is possible to send user data with that request. When using this particular subprofile between the MD-IS and MDBS, a correlation must be made between the TP4 connection itself and the channel stream being used to service the connection. Included in this user data field is information pertaining to the channel stream number being occupied, the cell number itself, the service provider's identifier, and a wide area service identifier, which is a marketing identifier for CDPD service providers (Fig. 3-12). What these terms mean, and

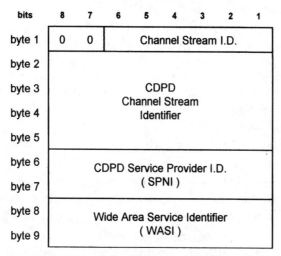

Figure 3-12 When using TP4 between MTSO and cell sites, RF channel attributes must be associated with each transport connection.

where they are applicable, will be elaborated on as we progress deeper into the workings of CDPD network elements.

As Fig. 3-11 indicates, the OSI TP4 protocol is supported by the OSI connectionless network service and a subnetwork-dependent convergence function which is responsible for data compression, encryption, and segmentation and reassembly processes. Other technologies may be used to deliver CDPD packets from a switching center (MD-IS) to remote cell sites, including X.25, frame relay, LAPD (another layer 2 hop-to-hop protocol), the very popular Point-to-Point Protocol (PPP), or DS0 (64 kilobytes per second) digital facilities. Use of DS0 channels is a common choice for delivering cellular voice traffic to mobile switching centers. For many carriers, it makes sense to use the same technology that is already in place for packet data delivery between their cell sites and switching centers.

Each of the disciplines in the areas marked with an arrow in Fig. 3-11 will be covered in detail in the following chapters.

Summary

- Each layer within a system provides a specific service. These services may possess connection-oriented (reliable) or connectionless (unreliable) attributes. While they are required at some layer within a stack, duplication of housekeeping activities such as sequencing, flow control, and error correction at many layers within a stack may adversely affect system performance.

■ Data is passed between layers within a system by the use of system calls known as primitives. Within a system, data is directed to various processes by the use of service access points. CDPD systems need to be aware of the SAPs that have been defined not only for the well-known standard protocols in use, but also for the CDPD specific functions such as mobile registration, authentication, and radio resource management.

■ The set of processes (layers) incorporated into a system is referred to as a profile.

■ CDPD elements incorporate three subprofiles: application subprofiles, lower layer subprofiles, and subnetwork subprofiles. Only end systems require application subprofiles.

■ Because each layer adds overhead and degrades the ratio of "payload" bits to total bits sent, CDPD attempts to regain some efficiencies by utilizing data compression and FEC algorithms.

■ Lower layer subprofiles have been defined for use among CDPD network elements for application and management or support traffic. CDPD carriers are free to choose subnetwork technologies that best meet their business needs.

4

CDPD Channels, Medium Access, and Control

Air Interface Profile

This chapter focuses on the bottom two layers of functionality between mobile database stations and mobile-end systems, the physical and the *medium access and control* (MAC) layers. Before doing so, however, let's observe how data is "filtered" down through the stack, from top to bottom. This will give us an idea of how the data is formatted before it is transmitted onto a cellular channel.

At the top of the air interface profile (Fig. 4-1) we see that, in addition to the IP and CLNP processes which transport user application data, there is a set of CDPD-specific entities which are unique to this mobile network. These new protocols are resident within the mobile-end systems themselves as well as the MD-IS elements. A chapter in this book will be devoted to the operation of each of these CDPD-specific processes, but at this time we will briefly define their roles in the system.

The *radio resource management protocol* (RRMP) is responsible for enabling the MDBS and M-ES equipment to cooperate with each other in the allocation and supervision of cellular radiofrequency use. Information relating to RF channel status and cell configuration is exchanged between mobile devices and the cell-site base station equipment. This information is used to determine "best" channels for initial CDPD use as well as what alternative channels should be used if a mobile device needs to hop to another channel because of contention, congestion, or channel degradation.

Airlink Protocol Profile

User Application Data	RRMP MNRP SMP	CDPD M - ES to MD - IS Operational Protocols	RRMP MNRP SMP	User Application Data
IP /CLNP		**Layer 3 CLNS**		IP / CLNP
SNDCP		**Encryption and Compression**		SNDCP
MDLP		**Hop to Hop Control**		MDLP
		MDLP Relay Function		
MAC		MAC	ISO 8043 TP4	The MD-IS to MDBS juncture may include the TP4 (shown) subprofile, or many other optional subprofiles including frame relay, PPP, LAPD and others.	ISO 8043 TP4
			CLNP		CLNP
			SNDCF		SNDCF
Physical		Physical	HDLC.....		HDLC.....
			DS0 - 802.3		DS0 - 802.3

M-ES   MDBS   MD-IS

CDPD - specific messages as well as User data are transported over the air interface.

Figure 4-1 Air link protocols.

Air link security is maintained by the *security management protocol* (SMP) and the *mobile network registration protocol* (MNRP). Together, they work to ensure that unauthorized access to the CDPD network is denied, and that once a mobile system is granted access, data transmissions over the air link are protected via encryption and/or compression. Specifically, the SMP is responsible for electronic key management procedures executed between M-ESs and MD-ISs. Each time a data link connection is established, the key sequences used for cyphering data change. The SMP provides a means to recalculate and exchange the new key sequences over the air link, thus protecting user authentication "credentials" from eavesdropping.

The MNRP is responsible for identifying a mobile device as a bonafide device that should be granted access to the network. This protocol provides for an exchange of credentials between mobile devices and the network. The credentials consist of a set of numbers that are initially programmed into an M-ES at subscription time, and subsequently changed each time a registration onto the network occurs. Some of this authentication information is changed sequentially, and some is changed according to a random number generator. The credentials of a mobile device change dynamically, and the "secret" codes that are created with each registration are shared by the M-ES and MD-ISs. Thus, mobile equipment and "home" MD-ISs of subscribers cooperate with each other to verify registration attempts as being made from an "authentic" device.

Cellular channel airspace is not by nature a "clean" medium for data transmission. There are, however, many ways to "hide" unreliable attributes of a communications medium from users of the channel. *Forward error correction* (FEC) and/or *automatic retransmission upon request* (ARQ) protocols are some commonly used mechanisms. Also, limiting the throughput of a communications channel via reduction of the signaling scheme's baud rate helps to minimize the number of errors encountered during a transmission. Often, data compression is used to compensate for low-speed serial link throughputs and the overhead introduced by the many protocols incorporated to assist with error detection and correction.

The next layer down is the *subnetwork-dependent convergence protocol* (SNDCP). It is here where encryption and compression of packet data occurs. The SNDCP attempts to make more efficient use of the air link by compressing IP packet headers according to the RFC1144 Van Jacobsen algorithm. CLNP headers are compressed using an approach that is similar to the Van Jacobsen algorithm in principal, but unique in that protocol fields with CLNP are obviously different than those in an IP header. Compression of user data being transported across the air interface is accomplished via the industry standard CCITT / ITU V.42bis. Segmentation of up to 2048-byte network layer packets and reassembly of 130-byte data link control data units also occur at this level. The segmentation and reassembly procedures performed by the SNDCP make better use of air link resources as opposed to having the IP/CLNP functions perform the same task. Obviously, the compression algorithms supported by the SNDCP also serve to make more efficient use of limited air link resources.

At the data link control level, we find the *mobile data link protocol* (MDLP). This is where the data link connections between mobile equipment and serving MD-ISs occur. A close relative of the ISDN *link access protocol for D channels* (LAPD), the MDLP provides a service that hides undesirables such as lost or missing frames from the higher layers. This protocol provides a connection-oriented delivery mechanism with full information sequencing, retransmission, and flow control capabilities. The maximum amount of (higher-layer) information that can be carried by an MDLP frame is 130 bytes. Network administrative information may be passed using an unacknowledged mode of operation, as opposed to the fully acknowledged information transfer associated with user data transfers.

Below the data link control level of the air interface is the medium access and control and physical (RF channel) layers, which are the focus of this chapter.

CDPD Channel Attributes

Cellular Industry Standards EIA/TIA-553, -54, -55, and -56 define issues and requirements for compatibility between mobile and fixed land stations. Radio channel assignments, frequencies, and power levels for CDPD equipment follow the recommendations set forth in these industry standards. What follows here is a brief digest of AMPS cellular channel characteristics. This discussion is not intended to be a comprehensive reference on issues relating to RF engineering, cellular or otherwise. Readers interested in topics related specifically to the cellular RF environment should consult these industry standards and other references listed in the bibliography.

Cellular channels are arranged in "pairs" consisting of a 30-kilohertz forward channel and a 30-kilohertz reverse channel. The forward channel is used for carrier-to-mobile transmissions, while the reverse channel is used for mobile-to-carrier transmissions. In order to minimize interference between the forward and reverse channels, they are separated from each other by 45 megahertz, as shown in Fig. 4-2. Channel 1 is the channel with an 870.030 forward frequency and an 825.030 reverse frequency.

A "basic" system includes 10 megahertz for cellular carrier A and 10 megahertz for cellular carrier B. Carrier A occupies channels 1 through 333. The corresponding frequencies for carrier A are 870.030 through 879.990 megahertz forward and 825.030 through 834.990 megahertz reverse (see Fig. 4-3). Carrier B operates on channels 334 through 666. The corresponding frequencies for carrier B are 880.020 through 889.980 megahertz forward and 835.020 through 844.980 megahertz reverse.

Each carrier may operate an optional "extended" system, known as A' and B'. The A' system frequencies occupy channel numbers 667 through 716. B' system frequencies occupy channel numbers 717 through 799. In addition, the A carrier may utilize another extended set of frequencies, known as A". The A" channel assignments are numbered 991 through 1023. The set of frequencies corresponding to channel numbers 800 through 990 are not allocated for cellular use.

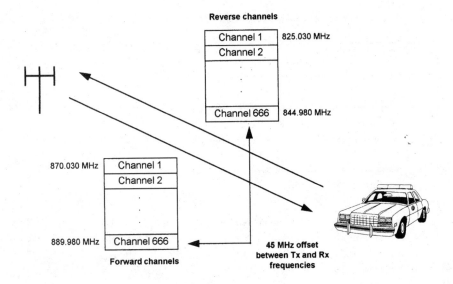

Basic cellular system frequencies

Figure 4-2 AMPS spectrum.

System	MHz	Total Channels	Boundary Channels	Forward Frequency (MHz)	Reverse Frequency (MHz)
Not used		1	990	824.010	842.010
A"	1	33	991 1023	824.040 825.000	869.040 870.000
A	10	333	1 333	825.030 834.990	870.030 879.990
B	10	333	334 666	835.020 844.980	880.020 889.980
A'	1.5	50	667 716	845.010 846.480	890.010 891.480
B'	2.5	83	717 799	846.510 848.970	891.510 893.970

Channel Numbers and Frequencies

Figure 4-3 Frequency allocations.

M - ES Power Level Classes - ERP

I	+ 6 dBW = 4.0 W
II	+2 dBW = 1.6 W
III	- 2 dBW = 0.6 W
IV	-2 dBW = 0.6 W

Figure 4-4   M-ES power classifications.

Although they are not directly applicable for CDPD use, the AMPS system has defined a set of control channels which occupy the frequencies of channels 313 through 354, 688 through 708, and 737 through 757. These control channels are used for initial channel acquisition, paging, and other mobile telephone service management procedures.

CDPD mobile equipment can be categorized into four equipment power classes, as indicated in Fig. 4-4. Output power tolerances should be between + 2 and −4 decibels of the nominal values indicated. Regardless of the power class of the mobile device, transmitter "ramp-up" and "ramp-down" time should be 2 milliseconds. Keeping these ramp times within 2 milliseconds is essential in order to minimize radio channel access times for mobile equipment sharing a common RF channel. Figure 4-5 shows that the cellular channel is essentially a point-to-multipoint access facility. The mobile database station acting as the central site operates its transmitter in *constant carrier* fashion. Of course, this constant forward channel output applies only to RF channels configured for CDPD and "in use" by CDPD. Recall that an RF channel can be shared by CDPD and circuit switched services. When a shared channel is in use by a non-CDPD service, the MDBS will be "quiet" as far as its forward channel output is concerned. All mobile devices sharing an in-use CDPD channel receive the MDBS's forward channel broadcast.

In the reverse direction, mobile devices must take turns accessing the facility. An M-ES will operate in what is referred to as *switched carrier* mode in the reverse direction. That is, when a transmission occurs, the mobile device turns on (ramps up) its carrier, modulates the data that needs to be sent, and then turns off (ramps down) its carrier. As is the case with multipoint landline systems, when a CDPD device is transmitting toward the "central site" (MDBS), no other remote users can access the channel. This is illustrated in

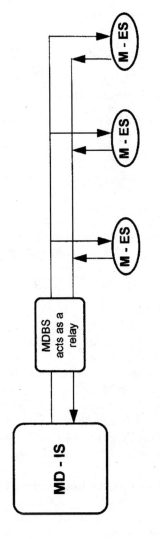

Logical view of CDPD channel space and airlink topology.

Figure 4-5 The air link as a multipoint network.

57

Accessing the CDPD Reverse Channel

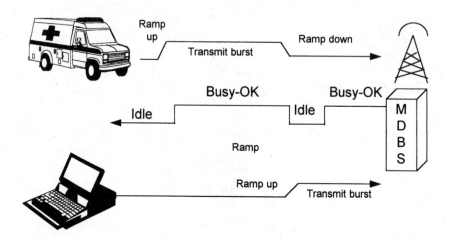

CDPD mobile - end systems operate on a point - to - multipoint facility and can output only after other stations on the same channel have ceased their transmissions.

Figure 4-6 Mobile-end systems must contend for reverse channel access.

Fig. 4-6, where the laptop computer must wait until the emergency vehicle has completed its transmission. Hence, rapid ramp-up and ramp-down times are essential for speedy access to the facility. In the event that an M-ES does not ramp down within a tolerable time frame, the malfunctioning mobile transmitter can be disabled by the CDPD carrier via a special ZAP command, which is designed into the system for this purpose. The CDPD ZAP command is analogous to antistreaming timers commonly implemented in multipoint-capable landline modems. Disabling of a CDPD "streamer" should be supported on an automated basis in order to ensure the availability of precious RF channel resources.

Unlike landline systems, however, polling sequences are not used as the mechanism by which CDPD mobile devices are granted access to the reverse channel. A unique medium access and control (MAC) process has been defined for this purpose.

Medium Access and Control

The CDPD MAC layer is closely related in principle to the operation of an 802.3 Ethernet-type local area network (LAN). In both environments, a station with a pending transmission examines the transport medium to see if it is idle or busy. An 802.3 LAN uses DC voltage levels detected on a wire as the busy/idle detect mechanism. Obviously, an M-ES cannot rely on DC voltage levels to be detected on a radiofrequency channel. Instead, a CDPD mobile-end system relies on logic contained within the forward channel stream broadcast by the MDBS, which indicates if the reverse channel stream is idle or busy. In both cases, the principle of waiting until an idle channel is detected prior to output is considered to be a "nondeterministic" access method. *Nondeterministic* implies that no token passing or polling mechanisms are employed by which an end system can determine that no other stations are using the facility at the present time. In an Ethernet environment, this nondeterministic access method is called *carrier sense multiple access* (CSMA). Because CDPD devices cannot sense DC voltage levels, they instead look for digital information in the form of bits that convey busy or idle status. We refer to the CDPD's access method as *digital sense multiple access* (DSMA).

It is entirely possible, however, that two end systems may be waiting for an idle channel at the same time. When the channel becomes available, the two end systems both begin transmission in the reverse direction. Because the channel is a shared medium, a collision between the two reverse transmission "bursts" occurs. In an Ethernet environment, voltage levels on the cable will rise to a level where the computers' network interface cards will be able to determine that a collision has occurred. Similarly, a CDPD mobile device needs to determine if a collision occurs on the shared reverse channel. In this event of a collision between two outputting M-ESs, CDPD's mobile database station receives two incoming signals that interfere with each other. Bit errors occur that cannot be compensated for via forward error correction procedures, and the resulting decode of the bit stream will be unsuccessful. The forward channel stream will indicate during "busy" conditions if the bit stream on the reverse channel was decoded successfully or not. As with the previously mentioned "busy/idle" indications, a series of bits is built into the CDPD forward channel stream's block structure that determine whether block decodes are successful or not. Transmitting CDPD mobile devices continuously monitor the forward channel stream for decode successful/unsuccessful information to determine if the transmission burst was received in error. In Fig. 4-6, note that the transmitting emergency vehicle is monitoring the forward channel stream as the reverse transmission burst continues. The forward channel is indicating "busy," preventing the laptop from accessing the reverse channel. In addition, the "OK" indicates that no bit errors are being detected. As a result, the emergency vehicle continues to transmit until it has completed its reverse channel burst. In the event that a transmitting M-ES determines that its output has been received in error (due to collisions or otherwise), the M-ES ceases transmission. The M-ES discovers this via reception of decode unsuccessful indicators imbedded into the forward channel stream. The M-ES then goes back to the busy/idle wait state, and retransmits the previous burst when an idle condition is again detected. This concept is called *collision detection* (CD). Accordingly, the set of rules governing CDPD mobile-end system access to a reverse channel is referred to as *digital sense multiple access with collision detection* (DSMA/CD).

The Forward Channel Stream

Figure 4-7 details the format of the forward channel block structure. The bit rate of the forward channel stream (as well as the reverse channel stream) is 19,200 bits per second. At the MAC layer, CDPD transmits information in the form of Reed Solomon blocks. At the top of Fig. 4-7, we see the format of a Reed Solomon 63,47 block. The numbers 63,47 indicate that out of 63 total symbols, 47 represent data; the difference is parity information used for FEC calculation. Reed Solomon codes are some of the most powerful forward error detection and correction mechanisms used in communications systems today, the mathematics of which is beyond the scope of this book. For the interested, a generic discussion of error control techniques and algorithms can be found in Odenwalder (1985) and in Lin and Costello (1983) (see Bibliography).

The CDPD forward channel block consists of a total of 70 symbols, each of which is 6 bits in length. Starting with the first symbol on the left, every tenth symbol is a "control flag" used for synchronization as well as busy/idle and decode status notification. Seven such control flags are found within a single block. These symbols are not considered in the Reed Solomon designation 63,47. They are overhead symbols required by the CDPD MAC protocol. Traditionally, protocols used over landlines provide synchronization information before each block commences in the form of HDLC "flag" sequences, or perhaps a string of synchronization characters, or "syns." Unlike these protocols, CDPD synchronization information is distributed at periodic intervals within each Reed Solomon block sent. Because the forward channel stream is a continuous broadcast, it is important to provide synchronization information continually at the MAC level for the M-ES population using a CDPD channel.

Below the representation of a single block is a breakdown of the seven control flags. Notice that the sixth position contains one bit of a 7-bit-long decode status flag. The decode status flag is set to 0000000, indicating that the previously received block on the reverse channel was decoded successfully by the MAC layer. A successful decode is one that passes the Reed Solomon parity checks and/or that contains bit

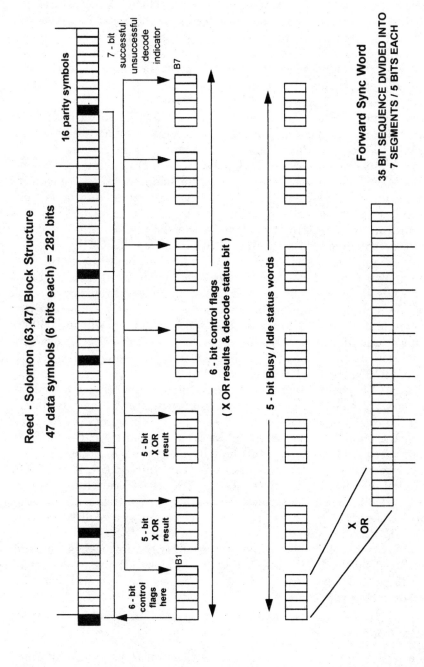

Figure 4-7 A forward channel MAC frame.

61

errors the FEC processes were able to correct. Conversely, a decode status flag of 1111111 indicates that the previously received block on the reverse channel contained bit errors that could not be corrected. Per CDPD R1.1, the actual number of symbols that can be corrected may vary with a vendor's implementation. It is suggested, however, that with the ability to correct seven symbols on a channel with a *co-channel interference* (C/I) ratio of 17 decibels, the undetected block error rate that can be expected will be approximately 1 block every 87 days. When a decode failure occurs, an M-ES will perform what is referred to as *exponential backoff,* in a manner similar to Ethernet's CSMA/CD access method, and attempt to reaccess the reverse channel and retransmit the errored block. The parameters affecting exponential backoff time intervals and retransmission counters are configurable within an M-ES. CDPD R1.1-recommended default values are given in Fig. 4-11 on page 67.

Continuing with our discussion of the forward channel stream, there are five more bits within each control flag. These bits are the result of an X-OR operation performed on a 5-bit busy/idle status flag and a 5-bit segment of a forward synchronization word that is a total of 35 bits in length. The parsing of the 35-bit synchronization word and Boolean operation on the busy/idle status flags is shown in the bottom half of Fig. 4-7. Busy is indicated by the binary string 00000, while idle is indicated by the binary string 11111. The 35-bit forward synchronization word, indicated in seven 5-bit sequences, is 11101 000001 11000 00100 11001 01010 01111.

The M-ES population observes the control flags being received on the forward channel stream to maintain synchronization and determines if the reverse channel may be accessed for a transmission burst toward the CDPD network. As an M-ES is transmitting, the forward channel control flags are continuously monitored to determine the integrity of transmitted bits as received by the MDBS. If an M-ES determines that an unsuccessful decode of its transmission has occurred, the M-ES immediately ceases its transmission. Unsuccessful decodes may occur due to noise, interference, or collisions with other M-ESs transmitting at the same time on the same channel— entirely possible when considering the random access method employed by the CDPD MAC. Retransmissions consist of the set of blocks needed to transmit any layer 2 frames that are still awaiting successful transmission.

Color Codes

The first 8 bits in each block transmitted on the forward channel stream contain a CDPD color code (Fig. 4-8). The code itself is broken down into a 3-bit field indicating an area color and a 5-bit field indi-

Color Code Encoding

8	7	6	5	4	3	2	1
Area Color			Cell Group Color				

All channels transmitted in a set of cells controlled by an individual MD-IS share the
same area color.

Figure 4-8 Color code identification.

cating a cell group color. The 3-bit area field is common to all channel
streams under the control of a given MD-IS. The 5-bit cell group color
is common to all cells belonging to a common cell group as long as
both of the following conditions apply:

1. The cell is adjacent to another cell in the same group.
2. All RF channels are used only once within a given group, regardless of how many cells are in the group.

   It is common practice to reuse RF channel frequencies when providing cellular coverage. One practice is to assign frequencies to channels within cells that are part of a common group. Each channel frequency is unique within a given group topology. A cell group may
consist of 7 cells, with any frequency being used only once within the
group. That topology, however, may be copied numerous times over
the *cellular geographic service area* (CGSA) of the carrier. Groups are
deployed in such a manner as to ensure that any given channel being
used within a cell group is physically separated by distance from a
cell using the same frequencies in another group. Figure 4-9 illustrates this point.
   Color codes are used to detect co-channel interference on frequencies in use by CDPD. An M-ES will detect the color codes being
received on the forward channel and echo these codes back toward the
MDBS on the reverse channel. If there is a conflict between color
codes being transmitted by the MDBS and color codes being transmitted by an M-ES, it can be assumed that these two devices operating
on a common frequency are being interfered with by a device operating on the same frequency, but transmitting from a different cell
group. If an MDBS detects a color code being received on a reverse
channel other than the codes being transmitted on the forward channel, the MDBS discards all received blocks and prevents any additional M-ES transmissions on that channel by declaring the channel busy.
Carrier-specific recovery procedures may apply at this point to determine when the channel may be put back into service. Similarly, if an

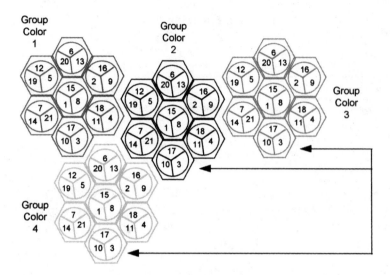

All cells are under the control of a single MD - IS (area).
Each group of 7 cells has its own group color.
Channels are re-used but separated by distance, with
many other channels being used in between.

Area Color

MD - IS

Cell Group Topology

Figure 4-9   Frequency reuse.

M-ES determines that color codes being received on the forward channel are suddenly different from the color codes currently being echoed back to the serving MDBS, the mobile device ceases transmission on the current channel and searches for another channel on which to continue transmission.

The Reverse Channel Stream

Figure 4-10 illustrates the MAC block structure utilized on the reverse channel. Notice that in the reverse direction, we have the concept of a channel "burst." A channel burst consists of a continuous transmission of a series of Reed Solomon blocks. The number of blocks transmitted is determined by the number of layer 2 frames that need to be sent based on current queue lengths. Mapping of bits from layer 2 frames into MAC blocks does not occur on a one-to-one basis. Recall that the

Dotting Sequence	Reverse Sync Word	First 63, 47 Block	Second Block	Next Block	Etc...	Ramp Down
38 bits	22 bits	378 bits (inc parity)	378 bits	378 bits	Until complete	2 ms max

REVERSE CHANNEL
Reed - Solomon (63,47) Block Structure

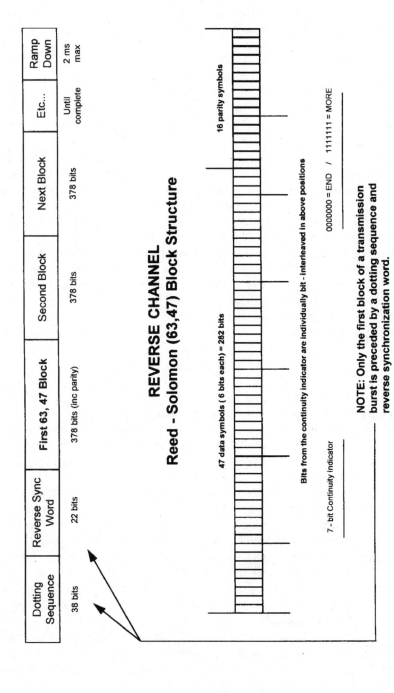

47 data symbols (6 bits each) = 282 bits

16 parity symbols

Bits from the continuity indicator are individually bit - interleaved in above positions

7 - bit Continuity Indicator

0000000 = END / 1111111 = MORE

NOTE: Only the first block of a transmission burst is preceded by a dotting sequence and reverse synchronization word.

Figure 4-10 A reverse channel MAC frame.

maximum information field at layer 2 over the CDPD air interface is 130 bytes. Together with addressing and control overhead, a layer 2 frame may approach 136 bytes in its entirety, 138 bytes if opening and closing HDLC flags are taken into account. Simple arithmetic tells us that a frame the size of 138×8 bits (1104 bits) is larger than a Reed Solomon block containing 47 data symbols or 282 bits. Accordingly, in order to transmit one full layer 2 CDPD frame, a transmission burst must consist of four CDPD MAC blocks. CDPD's layer 2 protocol utilizes sliding windows, and it is possible for up to 15 consecutive layer 2 frames to be sent in one transmission burst.

Transmission bursts in the reverse direction begin with a 38-bit dotting sequence of alternating ones and zeros (101010...), which serves as a training sequence for the MDBS receiver. The dotting sequence is sent during the M-ES transmitter's carrier ramp-up interval. Immediately following the dotting sequence is a 22-bit reverse synchronization word (1011 1011 0101 1001 1100 00) which allows the MDBS to delineate blocks and obtain synchronization with an M-ES.

The first entire 6-bit data symbol and the first 2 bits within the second 6-bit data symbol of an M-ES's transmission burst contain the color code being detected by the M-ES on the forward channel. While the MDBS includes the color code in the first 8 bits of each block transmitted, the M-ES includes the echoed color code only in the first 8 bits of the first block within a multiple block transmission burst. M-ES blocks are sent continuously until all blocks on queue have been transmitted.

Because the MDBS has no idea of when a transmission burst will end, an M-ES provides a continuity indicator within each block transmitted in the reverse direction. The continuity indicator advises the MDBS if additional blocks will follow in a transmission burst, or if a particular block is the last block to be sent within the current burst. The continuity indicator itself is a 7-bit word which is interleaved among data symbols within the Reed Solomon blocks. Specifically, 1 bit of the continuity indicator is found in every ninth 6-bit symbol. If the first symbol transmitted is considered s0, the first continuity indicator bit occupies the last position in symbol s7. The remaining continuity indicator bits are found just before the first bits in symbols s17, s26, s35, s44, s53, and s62. Figure 4-10 identifies the positions which are reserved for this purpose. After transmitting the last block within a transmission burst, the M-ES must ramp down its transmitter within 2.0 milliseconds. If the M-ES malfunctions and the transmitter fails to ramp down, the previously stated scenario of "zapping" the M-ES applies. When a ZAP command is actually sent to the mobile device that is streaming is determined by the CDPD service provider and equipment manufacturer.

Configurable M - ES MAC Parameters

Parameter	Default Value	Description
Max_TX_Attempts	13	The number of times an M-ES will observe Busy / Idle flags in order to gain access to a CDPD channel before declaring the channel congested.
Min_Idle_Time	0	The minimum amount of "microslots" an M-ES must remain in an Idle state before transmitting blocks. A microslot is the time between Busy / Idle flags, or 60 bits.
Max_Blocks	64	The maximum amount of blocks in one continuous transmission burst.
Max_Entrance_Delay	35	The maximum amount of microslots (60 bit times = 3.125 milliseconds) that the M-ES will wait when attempting to re-access a channel for an initial burst.
Min_Count	4	Due to decode failures, an M-ES will attempt to retransmit in no less than [2 (to the 4th power)] - 1 microslots time intervals.
Max_Count	8	Due to decode failures, an M-ES will attempt to retransmit in no more than [2 (to the 8th power)] - 1 microslots time intervals.

Figure 4-11 System parameters: RF media access.

MAC blocks are fixed in length, while layer 2 frames are variable in size up to a predefined maximum size of 136 bytes. It is possible that the last Reed Solomon block will need to be padded with interframe time fills. Interframe time fills may consist of flag sequences or a constant marking (binary 1) condition. All frames, however, must end with at least one flag sequence; if marks are being used as interframe time fill, marks should immediately follow the closing flag sequence. (See Fig. 4-11.)

5

CDPD's Mobile
Data Link Layer
Protocol

Responsibilities

The mobile data link protocol resides in layer 2 of the CDPD air link profile. This protocol provides a connection-oriented service with full housekeeping capabilities over the air interface. The housekeeping tasks performed by the MDLP work to create what appears to be a reliable communications channel for user/network datagram delivery. Because cellular channels are quite noisy, the duties of the MDLP are of paramount importance. Cellular channel space can be considered the point where most bit errors are expected to occur. The MDLP attempts to compensate for transmission errors over the hop that is most error prone. Rather than relying only on end system–to–end system error correction capabilities, MDLP contributes to increased network efficiency by correcting problems at the source. This chapter highlights the MDLP structure and elements of procedure.

Recall that the primary purpose of layer 2 data link control is to manage the transfer of information between network nodes, providing hop-to-hop control. Over the CDPD air interface, a logical hop exists between M-ES devices and the serving MD-IS. Note that the CDPD mobile database station is not included when discussing this logical hop. Generally speaking, we consider the MD-IS and the M-ESs endpoints of layer 2 connections, because these two entities are responsible for the framing of user datagrams. The MDBS equipment located between the M-ES population and a serving MD-IS simply relays network addressable messages over cellular channels and the physical

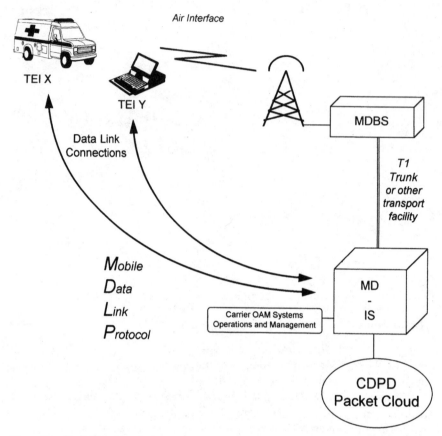

Figure 5-1 Air link hop control.

transport facilities that exist between cell sites and mobile switching centers (Fig. 5-1). In some instances base stations may originate messages toward mobile devices, specifically to manage radio resources such as adjusting an M-ES's output power. The MDBS, however, will never be the destination of M-ES-generated MDLP messages.

The CDPD mobile data link protocol is a close relative of the ISDN D channel protocol. Many of the commands, responses, and elements of procedures are identical to ISDN's LAPD, while some are specific to CDPD operation.

Message Structure

Figure 5-2 shows the format of an MDLP frame. At first glance, we see that the generic frame structure of [address field] [control field] [information field] is similar to many well-understood link level proto-

Mobile Data Link Protocol - MDLP Frame Format

Address Field	Control Field	INFORMATION FIELD Present only if control field indicates that this is an "information" frame.
1 to 4 Bytes	1 to 2 Bytes	130 Bytes Maximum Length

Figure 5-2 An MDLP frame.

cols already in widespread use. However, there is no link level "trailer" in the form of a frame check sequence (FCS). Accordingly, the link level control mechanism that CDPD employs over the air interface does not include any bit error detection mechanism. Instead, the Reed Solomon blocking found at the MAC layer is delegated responsibility for detecting and correcting on a forward basis any bit errors that may occur. This leads to a potential, and interesting, dilemma.

Suppose that CDPD services are provided in a manner similar to what is depicted in Fig. 5-3, where MDLP messages are simply relayed onto a DS0 transport system. Without additional encapsulation that includes an error detection scheme such as CRC checking, it is possible that bit errors could occur within the MDLP frame that would go undetected. Depending on the location within the frame where bit errors occur, MDLP address or control field information can change. If bits within the information field itself should change, it is possible that IP or CLNP checksums will detect these events. In the case where network level protocols detect these events, corrupt datagrams would be discarded. End system–to–end system error correction must now compensate for such network errors, perhaps in the form of a TCP time-out and retransmission. A carrier should employ a subnetwork subprofile between the MD-IS and MDBS that incorporates some error detection mechanism, such as LAPB, frame relay, or PPP. Recall the use of such optional subnetwork subprofiles between these particular network elements was suggested in Fig. 3-11.

MDLP Addressing

Our first major discussion of the MDLP focuses on the contents of the address field. CDPD modems are known to the serving MD-IS by an address called a *terminal equipment identifier* (TEI), also depicted in Fig. 5-1. TEI assignment is performed dynamically and automatically

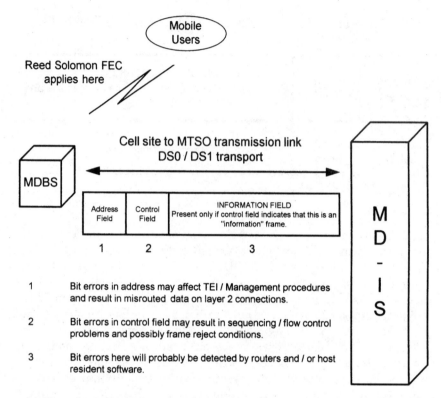

1   Bit errors in address may affect TEI / Management procedures and result in misrouted data on layer 2 connections.

2   Bit errors in control field may result in sequencing / flow control problems and possibly frame reject conditions.

3   Bit errors here will probably be detected by routers and / or host resident software.

Figure 5-3 MTSO-MDBS links require error control.

by the MD-IS which is serving the mobile device. Once a TEI has been assigned to a mobile device by the serving MD-IS, the data link connection between the two devices can be initialized. After initialization, the data link connection identifier is the TEI number itself. As Fig. 5-4 illustrates, the TEI may occupy as little as 6 bits of address space or as many as 27 bits. Consistent with other HDLC protocols, an address extension bit occupies the least significant position within each address byte. This bit is used to indicate if the byte that follows should also be interpreted as address information or control information, which happens to be the next field within the frame. A binary value of 0 indicates that the next byte is also part of the address field, which can be up to 4 bytes in total length. A binary value of 1 indicates that this is the last byte of addressing, and the next byte to follow is part of the control field.

The second most significant bit of the first address byte is used to indicate if the MDLP message being conveyed across the air interface is a logical command or response. Because the MDLP is connection

Mobile Data Link Protocol
MDLP Address Field

8	7	6	5	4	3	2	1
Temporary Equipment Identifier - TEI (MSBs)						C/R	E
May or may not be present							E
May or may not be present							E
TEI (LSBs)							1

TEI Values

0 = Layer 2 management procedures (such as TEI assignment and audits)
1 = Layer 3 broadcast services (such as channel configuration messages)
2 - 15 = Reserved
16 and up = Layer 3 point - to - point connections (such as M-ES to F - ES traffic)

The address field is a minimum of 1 byte, a maximum of 4 bytes. The field will be extended an additional byte as long as the extension bit indicates to do so via a value of "0". If the extension bit has a value of "1", then this is the last bit of the address field before the start of the control field.

Figure 5-4 MDLP addressing.

oriented, there are command verbs that require formal responses. Also, some verbs used within the MDLP vocabulary can be used as both command and response statements; the use of the C/R bit helps to minimize any ambiguity that may occur during link level dialogs. The manner in which this bit is used is consistent with both X.25's LAPB protocol and ISDN's LAPD and is depicted in Fig. 5-5.

TEI Assignment

The activities associated with TEI management can be categorized as TEI assignment, TEI audit, or TEI removal processes. All MDLP frames carrying TEI management messages utilize the broadcast TEI value 0 (refer to Fig. 5-4). We begin with the assignment process. After initial channel acquisition, a CDPD mobile-end system broadcasts a *TEI identity* (ID) *request* message to the serving MD-IS. Because this is a dynamically changing mobile environment and an MD-IS may need to process simultaneous multiple requests, there must be a means to identify an ID request message uniquely with a specific device in the field. This is accomplished by correlating a TEI request message with the *equipment identifier* (EID) of the requesting CDPD modem. Each CDPD device has a unique 48-bit number

MDLP Command Response Conventions

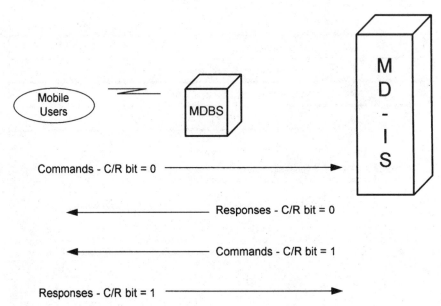

Figure 5-5 Command/response conventions.

assigned to it which is in the form of an IEEE MAC or hardware address. As is the case with common Ethernet network interface cards, EIDs are partitioned into a 3-byte manufacturer's product part and a 3-byte inventory number. (Refer to App. A.)

Also found within TEI ID request messages are parameters that are used to convey information relating to the desired link level attributes for this particular M-ES-to-MD-IS data link connection. Figure 5-6 illustrates the format of a TEI identity request message and shows the recommended values for each parameter according to R1.1 of the CDPD specification. Appendix B lists the format of all messages associated with the assignment, audit, and removal processes of TEI management.

Upon reception of a TEI ID request message from an M-ES, the serving MD-IS issues a *TEI assign* message that contains the assigned TEI as well as the link level attributes that are provisioned for this data link connection. In the event that system resources are not available to handle the original requested link level parameters, it is possible for the MD-IS to allocate lesser values in the TEI assign message. The format of the TEI assign message is indicated in Fig. 5-7.

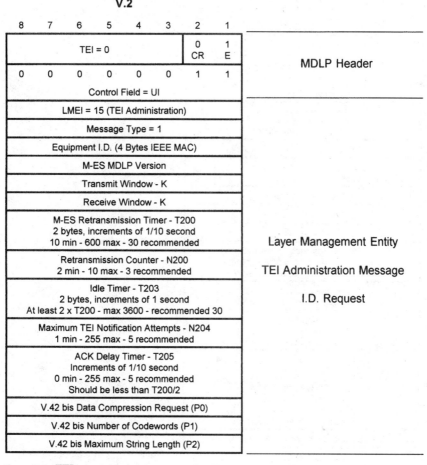

TEI Identity Request
V.2

8	7	6	5	4	3	2	1	
			TEI = 0			0 CR	1 E	MDLP Header
0	0	0	0	0	0	1	1	
			Control Field = UI					
LMEI = 15 (TEI Administration)								
Message Type = 1								
Equipment I.D. (4 Bytes IEEE MAC)								
M-ES MDLP Version								
Transmit Window - K								
Receive Window - K								
M-ES Retransmission Timer - T200 2 bytes, increments of 1/10 second 10 min - 600 max - 30 recommended								
Retransmission Counter - N200 2 min - 10 max - 3 recommended								Layer Management Entity
Idle Timer - T203 2 bytes, increments of 1 second At least 2 x T200 - max 3600 - recommended 30								TEI Administration Message
Maximum TEI Notification Attempts - N204 1 min - 255 max - 5 recommended								I.D. Request
ACK Delay Timer - T205 Increments of 1/10 second 0 min - 255 max - 5 recommended Should be less than T200/2								
V.42 bis Data Compression Request (P0)								
V.42 bis Number of Codewords (P1)								
V.42 bis Maximum String Length (P2)								

Figure 5-6   TEI request message.

TEI Audits and Removals

Audits of TEIs are accomplished via the *TEI check* message. TEI
checks are conducted when the network has a requirement to verify
that a TEI is indeed in use, or if there are multiple M-ESs being
served with the same, or duplicate TEI. Check messages are broadcast if there are no more available TEI addresses to assign, or if the
MD-IS has not heard from a given M-ES for an extended period of
time. One timer within the mobile CDPD device is a *configuration
timer,* which informs serving MD-ISs that an M-ES is available to
receive data. The default value for this timer is 4 hours. Accordingly,
if an MD-IS has not received any frames from an M-ES over a time

TEI Identity Assign
V.2

8	7	6	5	4	3	2	1
TEI = 0						1 CR	1 E
0	0	0	0	0	0	1	1
Control Field = UI							
LMEI = 15 (TEI Administration)							
Message Type = 2							
Equipment I.D. (4 Bytes IEEE MAC)							
ASSIGNED TEI VALUE Up to 4 bytes in length							Ext.
M-ES MDLP Version							
Transmit Window - K							
Receive Window - K							
M-ES Retransmission Timer - T200 2 bytes, increments of 1/10 second 10 min - 600 max - 30 recommended							
Retransmission Counter - N200 2 min - 10 max - 3 recommended							
Idle Timer - T203 2 bytes, increments of 1 second At least 2 x T200 - max 3600 - recommended 30							
Maximum TEI Notification Attempts - N204 1 min - 255 max - 5 recommended							
ACK Delay Timer - T205 Increments of 1/10 second 0 min - 255 max - 5 recommended Should be less than T200/2							
V.42 bis Data Compression Request (P0)							
V.42 bis Number of Codewords (P1)							
V.42 bis Maximum String Length (P2)							

Figure 5-7 TEI assign message.

period longer than 4 hours, a *TEI check request* message is transmitted to determine if the previously assigned TEI is still in use. An MD-IS transmitting a TEI check request message will wait "T201" seconds for a response. The recommended value for T201 is 5 seconds.

If no check response message is received within this time interval, a second check message is sent. If there is no reply message received after initiating a second request, the TEI value that is being checked is made available to other M-ESs requiring service. If a response to the request is received within this time interval, then the TEI in question is assumed to be in use. If two responses are received as a result of a single TEI check request, the EID contents are examined.

If the EID information in the multiple response messages is different, it is safe to say that the TEI value in question must be duplicated. In such an event, TEI removal procedures commence.

TEI removal is accomplished by an MD-IS issuing a *TEI remove* message. When an M-ES receives this message, all messages pending transmission are purged. An M-ES which has received a TEI remove message enters the TEI "unassigned" state and commences the procedures associated with acquiring an initial TEI from the serving MD-IS. At the carrier's end, TEIs that were removed are placed back into the available pool of TEIs for future use. TEI removal does not specifically require the transmission of a removal message toward an M-ES. A user's M-ES automatically stops using a given TEI and requests another TEI when entering another MD-IS's serving area. Other scenarios include when the MDLP enters a disconnect mode state, or when receiving an assignment message for a TEI that it is using, but the assignment message is being directed toward another equipment identifier.

MDLP Control

Immediately following the address field is the MDLP control field. As with most *high-level data link control* (HDLC) protocols, the control field defines the type of message being transferred, be it a link initialization verb, flow control verb, or even error notification.

Two modes of information transfer are associated with MDLP: unacknowledged information transfer and sequenced information transfer. Unacknowledged information transfer is used when a response is not required or is impractical to solicit. Management activities such as TEI assignment and auditing use this mode of information transfer because layer 2 confirmation is not required with these services. The management entities responsible for TEI inventories reside directly at either endpoint of the data link connection. If these entities receive messages, they respond with the appropriate return messages. If no responses are received, the management entities retransmit after the expiration of various timers. Hence, there is no need for a supporting layer such as MDLP to protect this type of information transfer with sequence numbering and retransmissions, because the layer immediately above assumes responsibility for managing its own activities.

With broadcast services, which include "limited," "directed," and "multicast" services, it is impractical to expect a response from multiple receivers, especially when considering the multipoint topology being used. Accordingly, these services make use of the unacknowledged information transfer capabilities of MDLP. All information transfer of this type includes a control field set to a binary value of 00000011, indicating a UI (unacknowledged information) frame. The templates we have seen thus far associated with TEI management messages bear this out.

When point-to-point transmission of user data occurs, the MDLP process utilizes the *sequenced information transfer mode* of operation. As the name implies, information sent over the air interface is accounted for by keeping track of sequence numbers which are part of the control field in each message. With this mode of operation, the link level protocol works to make sure that all messages are delivered in the same sequence that they were sent. Retransmissions of data link messages may occur in an effort to correct out-of-sequence frames or because no acknowledgments have been received. Flow control between the MD-IS and M-ES population is possible. In short, sequenced information transfer works toward providing guaranteed delivery of user data across point-to-point connections.

Figure 5-8 summarizes the set of MDLP commands and responses used to govern the air interface. Many of these verbs are identical to X.25's LAPB protocol and ISDN's LAPD, and they are used in the same manner. In addition to this set of what is perhaps well-understood verbs, TEST, ZAP, and "selective reject" have been added to CDPD's repertoire. A high-level discussion of these verbs and their use follows.

Mobile Data Link Protocol - MDLP Control Field			
FORMAT	TYPE COM or RESP	8 7 6 5 4 3 2 1	HEX CODING
Unnumbered	UI Command	0 0 0 0 0 0 1 1	03
	SABME Command	0 1 1 P 1 1 1 1	2F or 3F
	DISC Command	0 1 0 P 0 0 1 1	43 or 53
	UA Response	0 1 1 F 0 0 1 1	63 or 73
	DM Response	0 0 0 F 1 1 1 1	0F or 1F
	FRMR Response	1 0 0 F 0 1 1 1	87 or 97
	TEST Both	1 1 1 0 0 0 1 1	E3
	ZAP Command	1 1 1 0 1 1 1 1	EF
Supervisory	RR Both	0 0 0 0 0 0 0 1 NR Number PF	01 xx
	RNR Both	0 0 0 0 0 1 0 1 NR Number PF	05 xx
	Selective REJ Both	0 0 0 0 1 0 0 1 NR Number 0	09 xx
Information	INFO Command	NS Number 0 NR Number P	Even # xx

Figure 5-8   MDLP control.

Unnumbered Format Commands and Responses

A simple way to categorize the set of unnumbered commands and responses is to say that these verbs are used for mode setting and fault management purposes.

SABME.   SABME stands for *set asynchronous balanced mode with extended sequence numbering*. This command is used to initialize a data link connection between an M-ES which has had a TEI assigned and the serving MD-IS. The command also indicates that extended sequence numbering is desired for information transfer over this data link connection. Normal sequence numbering uses a modulus of 8, or sequence numbering from 0 through 7 on information frames transferred. (The initialization command is SABM.) Extended sequence numbering, provided via the SABME command, requests a modulus of 128, or sequence numbers in the range of 0 through 127 on information frames transmitted. Successful initialization of a data link with either command will enable two-way full duplex communications between participating endpoints. This mode of operation is referred to as *multiple frame establishment*.

DISC.   The *disconnect command* is used to release an existing association between an M-ES and serving MD-IS.

UA.   *Unnumbered acknowledgments* are affirmatives to previously received disconnect or SABME commands.

UI.   This control field is used for *unacknowledged information* transfer.

DM.   The *disconnect mode* response is generated by a layer 2 endpoint if the machine is already in a disconnected state and is unable to transfer information.

FRMR.   The *frame reject* response is generated when a link level abnormality occurs that cannot be corrected via retransmissions of frames. Abnormal conditions can cause this message to be generated. The message itself contains a diagnostic information field that helps to identify the cause of the event.

TEST.   This command is used to invoke a loopback test. Included in the message is an information field which is defined by the initiator of the loopback test.

ZAP.   This command is used for antistreaming purposes. When a malfunctioning M-ES fails to ramp down its transmitter after a transmission burst, this command can be used to disable the transmitter. The command has an information field of 1 to 4 bytes that indicates the number of seconds for which the targeted M-ES must shut its transmitter down. A transmitter can be enabled by sending another ZAP command with the information field indicating 0 seconds.

Supervisory Format Commands and Responses

RR. *Receiver ready* indicates that the endpoint issuing this verb is able to accept additional information frames starting with the indicated NR from the remote endpoint. An RR by itself is not an acknowledgment, but rather an indication of the state of the machine's receiver and its ability to handle additional information frames. In all cases, acknowledgment of information transfer is accomplished via an appropriate NR value.

RNR. *Receiver not ready* indicates to the remote endpoint that a temporary busy condition exists and the receiver is not able to accept additional information frames. The NR that is contained in the RNR verb indicates the next expected information frame that may be sent once the busy condition is cleared.

SREJ. *Selective reject* is a verb that a receiver can use to advise the remote endpoint that information which has been received was deemed to be out of sequence. Reception of an SREJ verb causes the transmitter of information frames to retransmit only the specific frame indicated by the SREJ's NR value.

Information Format Commands

INFO.   Information frames are always considered to be logical command statements. Unlike UI frames, INFO frames utilize sequence numbers which are used for acknowledgment purposes. "Now sending," or NS numbers, are incremented by the transmitter of INFO frames. Within each INFO frame is also a "next expected receive," or NR number, which refers to the next in sequence frame expected from the remote endpoint of the data link connection.

Poll/Final Bits

Notice that MDLP command verbs contain a bit marked P, while response verbs contain a bit marked F. Verbs that can be interpreted as either a command or a response have a bit that is designated PF. This bit is used for checkpointing purposes. In all cases, if this bit is set to a binary value of 0, then the bit has no significance whatsoever. However, if this bit is set to a binary value of 1, then the bit is being used as either a "poll" or a "final" indicator. Poll indication occurs in commands.

Final indication occurs in link level responses. Obviously, verbs that can be used as either a command or a response convey either a poll or a final indication, depending on the context in which the verb itself is used (for example, as either a command or a response).

As a general rule, a link level endpoint responds to commands that it receives. Data link entities allow a period of time within which a response must be received for previously issued commands. (A list of MDLP-associated timers and parameters can be found at the end of this chapter.) Initially, a command is issued with the P bit turned off. However, if the transmitter should time-out waiting for the expected response, it will retransmit the command with the P bit set to a binary 1. The same timer will be reset, and if it should expire once again, the command will again be sent with P = 1. This will continue up to a predetermined number of times or until the expected response is observed, but with the F bit set to a binary 1 condition. If no responses with a final bit are observed within the predefined number of attempts, reinitialization of the link occurs. Issuing commands with poll indications and ignoring responses until final indications are returned is referred to as *checkpointing*. Checking for final status when time-outs are occurring alleviates any ambiguity to those responses. An example of one such scenario is presented in Fig. 5-9.

Sequenced Information Transfer Operation

When sequenced information is being transferred, MDLP increments a 7-bit-long binary NS (send sequence number) within the control field. Sequencing commences with the number 0 immediately after a SABME/UA exchange. After a period of activity has elapsed, these sequence numbers will eventually reach the upper binary limit of 127. At this point, sequencing continues, starting again with the number 0. There is no requirement for reinitializing the data link when sequence numbers need to start over again from the 0. The numbers simply increment with each information frame that is transmitted, much the same way that an odometer increments mileage on a vehicle. No resets or other activity occurs when the odometer cycles from 999999 to 000000; the vehicle just keeps on going and the meter just keeps on turning. The NS, or "now sending" sequence number exists only within an information frame's control field, as indicated in Fig. 5-8. As a data link endpoint transmits information frames, this number is incremented.

An acknowledgment for any given information frame that is transmitted is the reception of another information frame or supervisory frame that has an NR number higher than the NS which was sent. An NR refers to the next (INFO) frame that a data link entity expects

Poll / Final Checkpointing

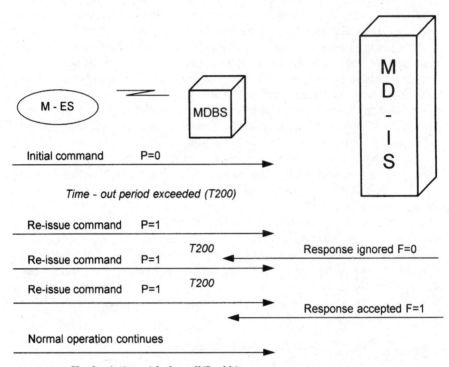

Figure 5-9 Checkpointing with the poll/final bit.

to receive from its remote peer. When receiving an NR from a remote endpoint, all information frames through (NR−1) are positively acknowledged. This acknowledgment (NR) can be found within supervisory RR, RNR, and SREJ frames as well as within INFO frames that are being transmitted from the remote endpoint.

MDLP does not utilize "negative acknowledgments," or *nak* verbs. Retransmissions occur automatically if acknowledgments have not been received for information that has been transmitted within a time period defined by MDLP system parameters. Retransmission of information may also occur due to reception of the supervisory SREJ verb. This verb is used to request a specific NS which was deemed to be missing within a multiple frame burst. The purpose of the SREJ verb is to resequence frames that arrive out of sequence.

MDLP is a windowing protocol. With such protocols, it is possible to transmit multiple frames in a single burst until either the first frame's T200 value has expired, or the amount of frames allowed to be transmitted within a given burst, called the k parameter, has been reached. If such a condition occurs, checkpointing commences in an attempt to solicit a response for the information frames sent previously. Acknowledgments can be "piggybacked," whereby one NR can represent a positive acknowledgment to all information frames received up until a value of NR−1. This is true when checkpointing and soliciting a response for previously transmitted frames, or during normal transmissions occurring within window and time-out restrictions. Figure 5-10 illustrates this scenario.

One interesting twist that MDLP incorporates into its link level repertoire that is not found in the popular LAPD and LAPB protocol is the use of timer T205. Within the LAPB and LAPD environments, it is common for a data link endpoint to respond immediately to infor-

Piggybacked Acknowledgments

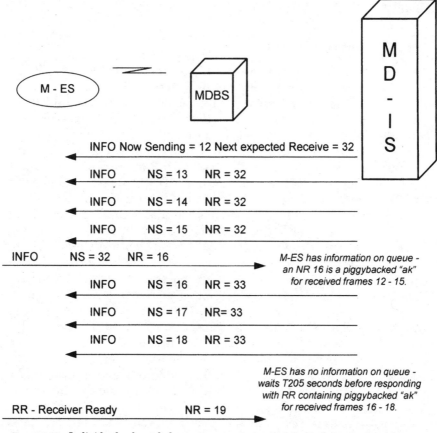

Figure 5-10   Individual acknowledgments are not required for each transmitted frame.

mation frames being received (see Fig. 5-12). When a receiving end-point has no information to output in the reverse direction, RR (receiver ready) responses are sent back toward the source of the information frames. Because the CDPD environment utilizes a multi-point topology, and also because of the random nature of the DSMA/CD access method, an M-ES incorporates the use of timer T205. This timer helps to ensure that the reverse channel stream is available for those mobile-end systems that have a need to output information, as opposed to allowing an acknowledging M-ES to monopolize the reverse channel bandwidth. T205 requires an end-point to wait a minimum of $\frac{1}{2}$ second before responding to informa-tion frames with the appropriate NR number imbedded in the super-visory RR frame. Doing so results in an M-ES issuing a single response to multiple frames instead of responding individually to each frame received. During that $\frac{1}{2}$-second pause it is possible that another M-ES on the same RF channel will wish to use the reverse channel; T205 gives other M-ESs that opportunity.

Corrupt or Missing Frames

As discussed earlier, MDLP does not perform data integrity checks on received frames. In the presence of bit errors, a few scenarios apply. The first and most desirable scenario is the case where bit errors occur on the RF segment, and Reed Solomon reliably detects and cor-rects the bit errors. The second scenario is a case that may even cre-ate unrecoverable link level problems. This would be the rare instance of a bit being changed in the MDLP address and/or control fields between the MDBS and serving MD-IS. A carrier deploying CDPD services would be ill-advised to not protect the MDBS-to-MD-IS interface with an encapsulation method that is capable of detecting errors. This is especially true when such interfaces require the use of terrestrial copper or microwave facilities, which are susceptible to transmission errors. Because transmission is the business of CDPD carriers, we can assume that this second scenario is not a concern: CDPD carriers most certainly will engineer remote MDBS-to-MD-IS interfaces in an appropriate manner. The third scenario is the case where Reed Solomon cannot compensate for errored bits on the RF segment, and appropriately discards the frame. The last scenario is the scenario where no frames are received by an endpoint because they are "lost" for some reason, perhaps due to signal dropouts. We will consider how the MDLP deals with the third and fourth scenar-ios, as there is a potential for these occurrences in actual practice.

Let us first consider what happens when only a single frame is transmitted, and that frame is not received at the remote endpoint.

Retransmission of single / last of burst INFO frames
Original INFO not received

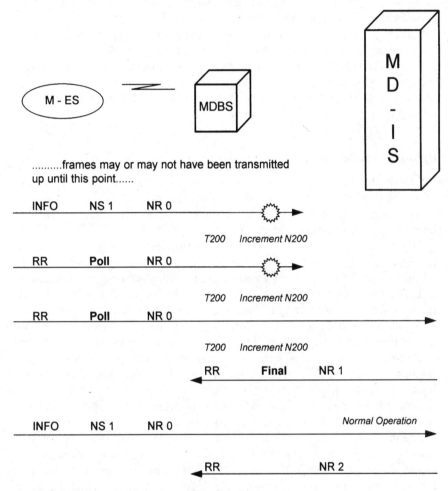

..........frames may or may not have been transmitted
up until this point......

INFO	NS 1	NR 0

T200   Increment N200

RR	**Poll**	NR 0

T200   Increment N200

RR	**Poll**	NR 0

T200   Increment N200

RR	**Final**	NR 1

INFO	NS 1	NR 0	*Normal Operation*

RR	NR 2

Figure 5-11   Recovery of information based on time-outs.

This can occur due to signal dropouts or a burst of bit errors that Reed
Solomon cannot correct. We will also assume that there is no flow control problem or busy condition. Figure 5-11 details the link level transactions that occur in such a scenario. The transmitting entity's T200
timer expires while waiting for an acknowledgment that never comes.
The transmitting endpoint then either transmits a supervisory RR
command with the poll bit set in an effort to solicit status from the
remote endpoint, or retransmits the information frame itself with the
poll bit set. Retransmissions of the RR or INFO message with poll bits

set will occur up to N200 times. If a supervisory frame with a final bit is not received within N200 retry attempts, link initialization commences and all information not acknowledged up to this time must be considered lost forever. This scenario also applies in the event that the last frame of a multiple-frame burst is not acknowledged within T200. Remember, however, that lost or missing information frames is not the only reason why this timer can expire. An information frame may travel across a data link, resulting in the desired response. It is possible, however, that the response itself may be lost because problems were encountered on the reverse path.

Using Selective Rejects

We have seen how MDLP recovers from a lost frame if the lost frame happens to be the last or the only frame transmitted. But what occurs if a missing frame is the first, or an intermediate frame within a burst consisting of many frames?

Most data link control mechanisms employ the use of a "go back and continue at N" retransmission scheme. This is accomplished via a REJect verb. When a burst consisting of multiple frames is transmitted, the originator commences retransmission starting with the NR number found within the REJect verb. The originator then continues transmitting any other frame that had been previously sent after the particular frame in question. The benefits of this approach is that "go back and continue at N" provides a simple way to ensure that all frames have been delivered in sequence. The downside of this technique is that entire "windows" worth of frames may have to be retransmitted even if only one frame was discarded or lost, as shown in Fig. 5-12. With large windows, a single discarded frame at or near the beginning of a transmission burst will cause multiple retransmissions, causing a degradation of throughput and response times on the affected link. Of course, this depends on whether the implementation finishes transmission of queued information frames before going back to retransmit, or if the implementation immediately aborts the current output and retransmits the rejected frame. Such variations in implementations do exist. This approach, shown in Fig. 5-12, works fine when the physical transport facility is known to be a reliable, error-free facility. Fiber-optic transport systems benefit from this mechanism because bit errors are rare, and large windows enhance transmission efficiencies. On a facility with frequent bit errors, this manner of retransmission will adversely affect overall network performance.

Because the cellular environment is prone to interference and bit errors can be expected, CDPD's data link protocol utilizes a selective reject approach to resequencing information transfer. Selective reject

"Go back and continue from N" retransmission

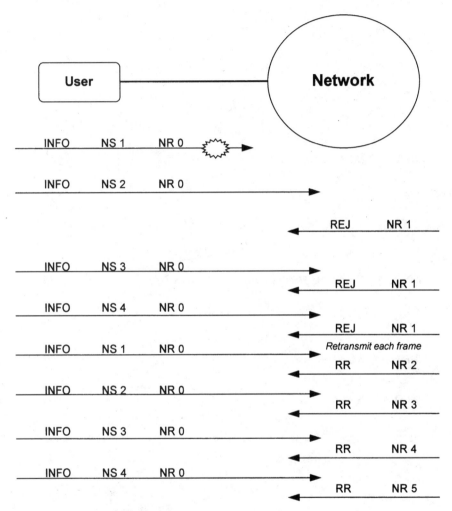

Figure 5-12 Using reject to recover information.

implementations can also be found over non-CDPD network links that have very large windows and/or long propagation delays, such as satellite links. The SREJ verb is used to request a retransmission of a specific frame, not "all frames starting with NR X." The benefit of this technique is that precious bandwidth is preserved by retransmitting only what is deemed to be missing. The downside associated with this method is that it is more difficult to implement in terms of software complexity and buffer management.

Selective Reject Example
Random lost frames

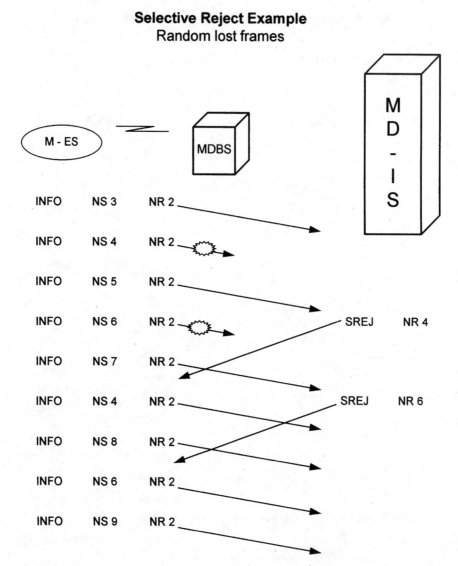

Figure 5-13 Selective reject in use.

Figure 5-13 provides an example of the use of selective reject to request retransmission of a single frame within a burst of many frames. As a second example, Fig. 5-14 represents a burst of consecutive missing frames and the use of SREJ to recover. Notice that an individual SREJ is issued for each individual NS that is deemed to be out of sequence. Also, the retransmission occurs immediately after notification that one is required, even though additional frames may

Selective Reject Example
Burst of lost frames

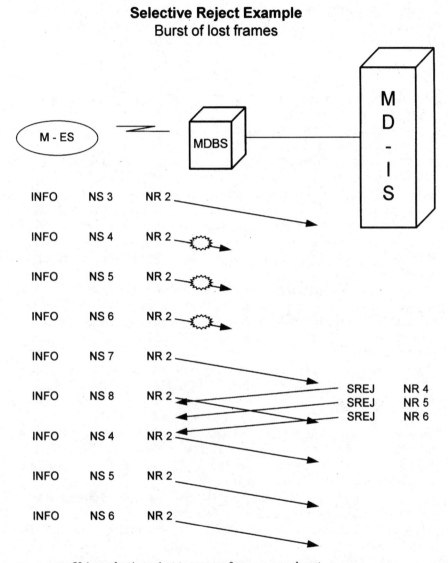

Figure 5-14   Using selective reject to recover from an error burst.

be in the transmit queue. There are many possible scenarios of either information or associated acknowledgments being missed. There are possibilities of the subsequent retransmissions themselves not being detected. Our purpose is not to cover every possible scenario; our objective is to understand that the use of selective reject support, regardless of the implementation, greatly enhances the performance of the CDPD air link by minimizing the amount of redundant information transmitted due to errors. At 19.2 kilobytes per second, any enhancements to transmission link efficiencies are highly desirable.

Sleep Mode

During the TEI request and assignment phases of M-ES operation, the mobile device has the option of specifying if *sleep mode* should apply for this data link connection. Sleep mode is requested by specifying a value within the T203 field of the TEI request message (refer to Fig. 5-6). If a value of 0 is defined, the M-ES is not electing to operate with sleep mode enabled. Any value specified in this field represents the length of time that an M-ES will remain fully powered while no frames of any type are being exchanged on its particular data link connection. When this value has been exceeded, the M-ES enters a power conservation state. The idea is to preserve battery capabilities, similar to power conservation efforts that portable laptop PCs incorporate.

Any time an M-ES has a need to transmit frames, sleep mode is abandoned and the mobile device operates normally, which is understandable. What may not be as easily understood is the manner in which messages can be sent to an M-ES which is sleeping at the moment.

Each channel stream has a notification timer associated with it. This timer, T204, defines the amount of time between TEI notification messages that are broadcast from the MD-IS toward the mobile population (Fig. 5-15). This message contains a list of TEIs that have information on queue which is destined for them. By default, these messages are broadcast every 60 seconds. M-ESs in the field must be able to discover the frequency of this list's distribution. Accordingly, the list is always being broadcast at the appropriate interval, even if the list is empty because no M-ESs have forward channel traffic pending. Within the notification message we see the current value of T204. By broadcasting current T204 values, mobile terminals are made aware of when the next notification will take place. This enables an M-ES to synchronize to the frequency of list distribution and periodically "wake up" to determine if any traffic is on queue for delivery.

When an M-ES determines that there is information on queue waiting to be delivered, a supervisory RR with the poll bit set is issued back toward the MD-IS. The MD-IS responds with an RR/final, followed by the information waiting to be delivered. If the target M-ES is not able to receive incoming information at the moment, an RNR frame is sent back toward the network. Another variable, N204, defines how many times an MD-IS will attempt to notify a mobile

TEI Notification Messages

8	7	6	5	4	3	2	1

						1 CR	1 E
TEI = 0							

0	0	0	0	0	0	1	1
Control Field = UI							

LMEI = 1 (Sleep - mode management)
Message type = 1 (TEI notification)
TEI notification timer in seconds

First TEI	Ext.
Remainder of list continues until complete	Ext.

Figure 5-15 TEI notification assists in managing "sleeping" mobile systems.

user of information pending transmission. What occurs when a TEI has reached its limit of notification attempts is implementation dependent. The recommended value for N204 is five attempts.

Frame Reject Conditions

MDLP incorporates frame reject (FRMR) support. Recall that a frame reject is a response that is generated due to abnormal conditions that cannot be compensated for by retransmitting frames. Therefore, it is important for any HDLC implementation to be able to trap and store these frames for later investigation, because normal procedure upon a frame reject event is to reinitialize the link level connection. Reinitialization causes all unacknowledged information to be discarded, frustrating users and perhaps disturbing applications. Many off-the-shelf HDLC products have very simplistic "management" areas within which counts of layer 2 frame types are kept. With respect to FRMR responses, it is desirable not only to count these events but also to store the associated diagnostic information, timestamp included, in a buffer. Doing so gives the system operator a means to track the cause of FRMR events. The diagnostic area provides valuable information that can be used to identify and correct problems causing link level reinitialization. Possible factors that can cause a FRMR event are shown in Fig. 5-16, which illustrates the format of a frame reject response frame.

MDLP Address Field (1 to 4 bytes)	TEI
Control Field (87 or 97 depending upon F bit)	1 0 0 F 0 1 1 1
	Control field of rejected frame (2 bytes)

Information / Diagnostic Field	V(S) 0
	V(R) C/R
	0 0 0 Z Y X W

If an unnumbered information frame, the second byte in the "rejected frame" field will be all zeros.

V (S) - The current send state variable of the station issuing the FRMR.

V (R) - The current receive state variable of the station issuing the FRMR.

C / R - If the rejected frame was considered to be a command, this bit will be set to a zero.

Z - Indicates the rejected frame's NR value was invalid. Check first two diagnostic bytes to observe.

Y - N201 exceeded. N201 represents the maximum size of the MDLP information field. (130 bytes)

X - The frame indicated in the first two diagnostic bytes was considered to be invalid because an information field was received which was not allowed for the indicated frame type. May also refer to an unnumbered or supervisory frame which was not the correct length, i.e., not a multiple of 8 bits or too few / too many octets. Bit W should be set in conjunction with this bit.

W - Indicates that the control field indicated in the first two diagnostic bytes was considered to be invalid, of a type that is not implemented, or undefined.

Figure 5-16 Unrecoverable link level problems result in frame reject generation. FRMR messages contain valuable diagnostics that can aid in identifying the cause of the problem.

MDLP System Timers and Parameters

T200 retransmission timer: The time interval that a transmitting data link entity will wait for a response to an issued command. Recommended default values are 3 seconds for user M-ESs and 5 seconds for MD-ISs.

N200 retransmission counter: The number of times that a command will be reissued with the P bit set. Reissued commands occur at T200-second intervals. The recommended default value is 3. If no response is received after N200 retries, link initialization procedures begin.

T201 TEI identity check timer: The minimum amount of time between MD-IS-generated TEI identity check messages. Issued when TEI resources have been exhausted or upon expiration of an M-ES's configuration timer, which by default is 4 hours.

N201 maximum information field size: The maximum number of bytes allowable in an MDLP information field. The system default, which cannot be modified, is 130 bytes.

T202 TEI identity request timer: The length of time between TEI identity request messages. This value is not configurable; the default setting is 5 seconds.

N202 TEI ID request retransmission timer: The number of times an M-ES will issue a request for a TEI to be assigned. The default value is three attempts.

T203 idle timer: The length of time an M-ES will allow a dateline connection to remain idle before it enters sleep mode. A value of 0 indicates that sleep mode is not desired. The recommended value is 30 seconds.

T204 TEI notification timer: The number of seconds until the next TEI notification is issued. TEI notification is used to advise sleeping M-ESs that information is on queue to be delivered. This value is also contained within the TEI notification message itself. The recommended default is 60 seconds.

N204 TEI notification counter: The length of time for which an MD-IS will attempt to notify a sleeping M-ES that information is on queue to be delivered. The recommended default is five attempts.

T205 acknowledgment delay timer: The minimum length of time a data link entity must wait before RR frames are transmitted to respond to information frames that have been received. This timer is implemented to ensure that a minimum amount of responses is issued for multiple I frame reception. The delay ensures that channel availability remains high.

K MDLP window size: The number of information frames that can be sent before an acknowledgment must be received. The parameter applies individually for both forward and reverse channels. The recommended value is 15 for both directions of information flow.

6

CDPD's Subnetwork-Dependent Convergence Protocol

Purpose

Just above MDLP and just below the network-level routing functions is the *subnetwork-dependent convergence protocol* (SNDCP). This protocol is responsible for ensuring the privacy of data sent over the air interface and resides within the MD-IS and M-ES entities. Three processes assist in accomplishing this. SNDCP first compresses the headers of the network-level routing protocol. After header compression is performed, datagrams are passed through a V42bis compressor. Lastly, user messages are segmented down to a size appropriate for transmission over the air interface and passed through an encryption process which is applied to payload data. The compression routines not only help make MDLP payloads secure, they offset inefficiencies caused by the various layering processes and relatively low channel speeds of 19,200 bits per second. The term *MDLP payload* is used because the information field at layer 2 is comprised of an SNDCP PDU. (Recall the discussion of SDUs and PDUs in Chap. 3). The maximum user data packet size that the segmentation and reassembly process will accept is 2048 bytes. Accordingly, a single SNDCP SDU may be presented to the mobile data link layer in as many as sixteen 130-byte-long PDUs.

SNDCP supports two classes of service. One class of service operates over the unacknowledged information transfer mode of MDLP. This mode of operation applies to broadcast services such as TEI management. The second class of service therefore must be fully sequenced information transfer. Recall that sequenced information transfer is applicable for point-to-point data link connections between an MD-IS and an M-ES. Figure 6-1 illustrates the functionality pro-

SNDCP Classes of Service

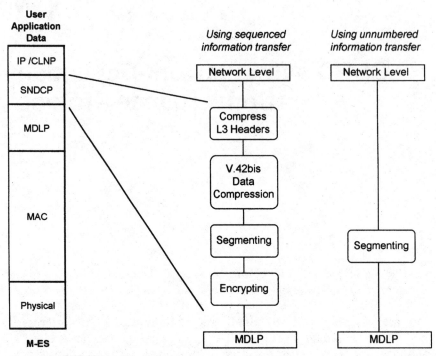

Figure 6-1 SNDCP services can vary.

vided for each class of SNDCP service. While the graphic shows an
M-ES, keep in mind that the processes indicated also apply to MD-
ISs. Because of the use of two modes of operation, the header or protocol control (PCI) information supplied by the SNDCP will vary
depending on the class of service used. Figure 6-2 illustrates both
header formats and also shows the location within user network messages where these headers may be found.

Segmentation Functions

As with any protocol that supports segmentation and reassembly procedures, there must be a mechanism to identify when a reassembled
PDU may be passed up the receiver's protocol stack. Only after the
last segment within a multiple segment stream of PDUs has been
identified can this occur. While many systems utilize individual "first
segment," "intermediate segment," and "last segment" indicators, the
SNDCP relies on a simple "more" segments to follow indicator.

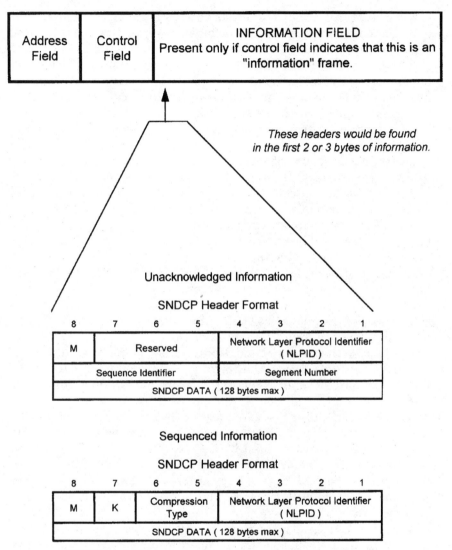

Figure 6-2 SNDCP headers.

Similar in approach to X.25's reassembly process, the SNDCP incorporates use of an M bit, or "more segments" indicator, within its header. When a receiver is informed that "no more" segments will follow, the PDUs that have been received are delivered to the layer above. A binary setting of 1 indicates that more PDUs will follow. A binary setting of 0 indicates that this is the last (or only) segment, and an SNDCP entity may pass what has been received thus far up the stack toward the network layer.

Notice that the M indicator exists with both sequenced information and unnumbered information transfer. However, with the unacknowledged mode of information transfer there are two additional fields, indicating sequence identification and segment number. Because no housekeeping chores take place at the link level with UI frame transmission, there is no guarantee that frames will be delivered. By using a segment number, the SNDCP is able to check whether all transmitted segments have been accounted for. An error exists if the receiver should determine after reception of a "no more" indicator (M = 0) that a noncontiguous stream of segments has been received. If this occurs, the receiver discards the incomplete set of PDUs, as the original stream cannot be reassembled. Multiplexing of multiple packet level associations is possible. However, over a given data link connection, no contiguous sequence of segments for an original SNDCP SDU may be interrupted by other SNDCP PDUs. In order to correlate a complete SNDCP PDU stream with a given layer 3 connection, a sequence identifier is used. Completely reassembled PDU streams are handed to the appropriate layer 3 process on a per-sequence identifier basis.

With respect to fully acknowledged information transfer, there is no need for sequence identifiers or segment numbers at the SNDCP level. Because the supporting MDLP process ensures that all frames are delivered in sequence, the convergence protocol can be streamlined by simply checking M-bit status before forwarding segments to the network level.

What network level processes are being served by the SNDCP? Recall the discussion in Chap. 3 of service access points, ports, and sockets. The mechanism used by the SNDCP to identify a layer 3 process is called a *network-level protocol identifier* (NLPID). Figure 6-3 identifies the NLPIDs used by the SNDCP. This value is part of the headers within each message. Figure 6-4 illustrates the reassembly process at work.

A set of TCP/IP headers without any options included consists of 40 bytes. One of the reasons that segmentation is performed by the SNDCP processes instead of relying on the network level process is that copying IP headers onto multiple messages creates an undesired amount of overhead which is passed over the air link. SNDCP's procedure of holding a segment until being advised that there are no more to follow is not only simpler, it provides higher efficiencies over the CDPD channel stream. By using the SNDCP to segment and reassemble, IP headers do not need to be copied to each message passed over the air interface. Instead, only the first frame, which carries a segmented IP datagram, carries the associated header information, as shown in Fig. 6-5. The receiving device's SNDCP reassembles

SNDCP
Network Layer Protocol Identifiers

Value	Meaning
0	MNRP Mobile Network Registration Protocol
1	SMP Security Management Entity Protocol
2	CLNP ISO 8473 Connectionless Network Protocol
3	IP DOD RFC 791 Internet Protocol
4 - 15	Reserved

Figure 6-3 Network layer protocol identifiers used by SNDCP.

Network Level Entities

MNRP 0	SMP 1	CLNP 2	IP 3

A SID Z Seg 1 M = 1 NLPID 2   Hold A

B SID Y Seg 1 M = 0 NLPID 3   Deliver B to IP

C SID X Seg 1 M = 1 NLPID 3   Hold C
D SID X Seg 2 M = 1 NLPID 3   Hold D
E SID X Seg 3 M = 0 NLPID 3   Deliver CDE to IP

F SID Z Seg 1 M = 1 NLPID 2   Deliver AF to CLNP

1 NLPID 1 M = 1   Hold
2 NLPID 1 M = 1   Hold
3 NLPID 1 M = 0   Deliver 1,2,3 to IP
4 NLPID 0 M = 0   Deliver to MNRP
5 NLPID 1 M = 0   Deliver to SMP

Subnetwork - Dependent Convergence Process

Unacknowledged Data

In-Sequence Acknowledged data

Mobile Data Link Protocol

Figure 6-4 Reassembling SNDCP PDUs.

An IP datagram of 1064 bytes is handed to SNDCP.

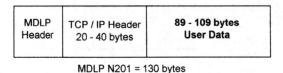

Relying on higher - level protocol for segmentation purposes degrades efficiency. First segment contains TCP/IP headers, remaining segments contain IP headers.

MDLP Header	TCP / IP Header 20 - 40 bytes	**89 - 109 bytes User Data**

MDLP N201 = 130 bytes

Relying on SNDCP for segmentation requires the transmission of the original header, and reliance on 1 - byte SNDCP header in subsequent segments results in more efficient data / overhead ratios.

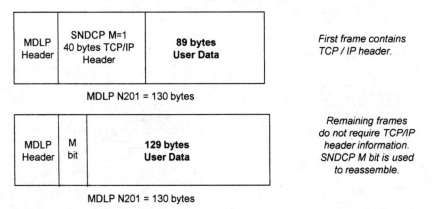

MDLP N201 = 130 bytes

MDLP N201 = 130 bytes

Figure 6-5 Segmentation and reassembly of IP datagrams can be handled more efficiently by SNDCP than by IP itself.

information into a complete SNDCP SDU and passes the IP datagram to the receiver's network layer.

The K bit shown in Fig. 6-2 refers to an encryption key sequence number. This bit alternates from a 0 to a 1 state after various stages of electronic key exchanges have been completed. When this bit changes state, a new electronic key is used for encryption purposes.

SNDCP Header Compression

In addition to "buying back" air link efficiencies by performing the segmentation function, the SNDCP also reduces the amount of data transmitted via compression algorithms. Recall that two levels of compression exist within the SNDCP. The first level consists of network-level header compression, while the second level consists of network-level data compression. The specifics of the attributes of user data compression are indicated in the TEI identity request and assignment phases of M-ES operation. (Recall the V.42bis parameter fields.) Network-level header compression specifics are identified within the acknowledged mode SNDCP header's "compression type" field. The compression algorithms vary with the network layer protocol in use, e.g., CLNP or IP.

Systems employing the TCP/IP defined in Internet RFC791 and RFC793 utilize the Van Jacobsen algorithms specified in Internet RFC1144. RFC1144 is a well-known set of definitions for compressing TCP and IP headers. The premise behind TCP/IP header compression is fairly straightforward. While the headers may consist of a total of 40 bytes, the intelligence within the header may need to be transmitted only once during the lifetime of a connection. It makes sense to build tables that remember the state of a given protocol field instead of redundantly transmitting a protocol field that has not changed since the previous datagram. If a given protocol field changes, then that intelligence should be relayed in the next compressed datagram header.

An example of this idea at work is to assign a connection identifier to a route between two communicating systems instead of relaying IP address pairs within each datagram. IP addresses are 4 bytes in length. A TCP/IP compressor can "remember" a connection between IP addresses and replace an 8-byte to/from address pair with a 1-byte connection identifier, shaving 7 bytes from each datagram associated with that connection. Another example is to transmit the delta between two sequence numbers instead of transmitting all of the sequence numbers themselves. RFC1144 looks at each field within a TCP/IP header and transmits only what needs to be transmitted in any given datagram. During an active TCP connection, it is possible to compress TCP/IP headers by omitting redundancies and substituting smaller fields for larger fields that have small magnitudes of change into as few as 3 bytes. The above scenarios are offered as examples to assist in appreciating RFC1144 behavior. The reader should read the RFC itself for complete details on the procedures for TCP/IP header compression. Refer to App. C for assistance.

Observe the compression type field within the SNDCP header. Three values apply for IP network-level operation. Binary 00 indicates that an IP header exists, but not a TCP header. This is possible because other processes besides TCP may reside above IP. Many proprietary network products make use of IP at the network level for routing purposes only. This was mentioned in Chap. 3 and shown in Fig. 3-2. Because RFC1144 operates on TCP as well, if the SNDCP header indicates 00/IP only, then no compression is performed on these datagrams. Binary 01 in this field indicates that there is an IP as well as a TCP header within the particular segment, but they are not compressed. Because TCP is connection oriented, the first and last segments exchanged between end systems are used for connection establishment and release procedures. In TCP terms, this is called *synchronization*. The intelligence conveyed during synchronization processes cannot be omitted, so these datagrams are not compressed. Subsequent datagrams on the established connections, however, are compressed. These compressed datagrams are identified via a binary value of 10 in SNDCP's compression type field.

Figure 6-6 shows the format of a compressed TCP/IP datagram. The fields indicated in *italics* are transmitted only if they differ from the values in the previous datagram. While these fields will be defined in this text, a familiarity with the IP and TCP is a prerequisite for understanding the relationship between uncompressed and compressed TCP/IP. The first byte in Fig. 6-6 is a change mask which identifies which fields are present within the compressed headers. A binary value of 1 in a field within the change mask indicates that the associated field has changed. Because there has been a change in the original datagram, the receiver parses the compressed header, takes note of the applicable field's new value, and adds the new set of parameters to its lookup table. A value of 0 indicates that the parameters

Compressed TCP / IP Header

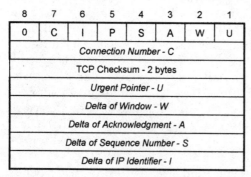

Figure 6-6 TCP/IP compression eliminates redundancies in transmitted datagrams.

in the identified field have not changed, and so are not included in the compressed header. Accordingly, the receiver uses whatever parameters are already resident within the compressor lookup table. Remember that a field is present only if there has been a change since the last datagram. Each of the italicized fields below the change mask correlates to the C, I, S, A, W, and U bits within the mask itself. An exception is the P bit, which is a copy of the "push" bit found in a normal TCP header. To protect data integrity, the TCP checksum is always transported within compressed datagrams.

The ISO 8473 connectionless network protocol may also be serviced by SNDCP. When applying compression to CLNP, the compression type field shown in Fig. 6-2 is set to the same values for CLNP only, uncompressed CLNP, and compressed CLNP datagrams that we have seen for TCP/IP. Compressed CLNP varies from compressed TCP/IP in that no compression is performed at the transport level; only the CLNP header is compressed. The principle of operation behind compressing CLNP headers is similar to that of RFC1144 in that redundancies among datagrams are omitted. A CLNP header may exceed 57 bytes; obviously, applying header compression to this protocol will result in better link utilization. The amount of redundant or nonsignificant data that can be eliminated is significant. According to CDPDR1.1, a 57-byte CLNP header can be reduced to as little as a single byte on PDUs following the initial L3 PDU in a data stream. A change mask similar to that used with TCP/IP compression identifies what fields, if any, have changed from the previous datagram.

Figure 6-7 illustrates an uncompressed CLNP header. As was the case with TCP/IP header compression, an understanding of the original protocol's operation, in this case ISO 8473 CLNP, helps when observing how it is compressed. Again, App. C in this book can be used as a guide to finding relevant information. Definition of each field within the CLNP header follows.

NLPID: Network-layer protocol ID / 81 = CLNP 8473

Header length indicator: Indicated in bytes

Version: Self-explanatory

Lifetime: Hops left to live before discard

SP: Segmentation permitted / 1 = yes / 0 = no

MS: More segments to follow / 1 = yes / 0 = last or only segment

ER: Error report generation on discard / 1 = yes / 0 = no

Type: Identifies type of datagram / 11100 = Data PDU

Segment length: Number of bytes in this segment

Header checksum: Self-explanatory

CNLP Header Format

8	7	6	5	4	3	2	1
NLPID - 81							
Header Length Indicator							
Version							
Lifetime							
SP	MS	ER	Type				
Segment Length							
Header Checksum							
Destination Address Length							
Destination Address							
Source Address Length							
Source Address							
Data Unit Identifier							
Segment Offset							
Total Length							
Options							

Figure 6-7 An uncompressed CLNP header.

Destination / Source address length fields: Indicated in number of bytes

Destination / Source addresses: Self-explanatory

Data unit identifier: "Tag" associated with original PDU before segmentation

Segment offset: How many bytes to displace this segment in order to reassemble

Total length: length of original PDU before segmentation

Options: Not always used; may include quality of service parameters

Italics above denote fields that would not change within a stream of datagrams being exchanged between two end systems. This redun-

Compressed CLNP Header

8	7	6	5	4	3	2	1
0	C	I	E	M	S	L	H

Address pair index - C
Delta of data unit identfier - I
Segment offset - S
Total length - L
Header length - H
Options

E and M fields correspond to the
Error Report and More Segment indicators.

Figure 6-8 As in TCP/IP compression, a change mask is used with CLNP header compression.

dant information need not be included in each datagram exchanged. Figure 6-8 indicates the change mask that is associated with CLNP compression. In this figure, *italics* are used for fields that may or may not change during an association, requiring the use of a mask.

V.42bis Data Compression

V.42bis compression attempts to substitute common occurrences of character strings with a fixed-length codeword. Codewords and the character strings they represent are stored in "dictionaries" that are dynamically updated as data exchanges occur. Both endpoints of a data link connection employing V.42bis operation must cooperate in maintaining identical dictionaries. This cooperation not only includes dictionary updates in the form of new character strings and codewords, but deletions of infrequently used character strings and codewords.

Figure 6-9 represents a block diagram of a V.42bis machine. An M-ES will physically connect to an asynchronous DTE via the set of interchange circuits provided, most likely RS232 compatible. Data is then transferred to a control function which manages the operational mode of the machine. V.42bis data compression applies only to sequenced, acknowledged information transfer, regardless of the layer 3 networking protocol implemented. Information transmitted over the air interface with the broadcast TEI or within unnumbered information (UI) frames is not compressed. Therefore, an obvious responsibility of the control function is to establish a mode where data is com-

V.42*bis* Machine

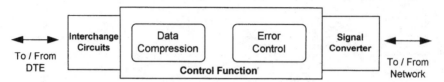

Figure 6-9 A block diagram of V.42bis error compression functions.

pressed, or passed transparently with no compression. If data compression is to be invoked, a set of negotiable parameters must be agreed upon by the communicating DCE devices. In our case, the communicating DCE devices are the MD-IS and M-ESs.

V.42bis requires the use of an error correction mechanism in order to ensure the proper operation of the compression and decompression processes. According to the ITU-T recommendation, LAPM (link access procedure for modems, V.42) or V.120 error correcting procedures should be supported. Both of these protocols provide for sequenced information transfer, in a manner similar to CDPD's link level MDLP. However, they also include the use of a frame check sequence of up to either 16 or 32 bits, which MDLP does not support. Therefore, V.42bis operation within the CDPD environment incorporates error control procedures unique to CDPD, MDLP frame transfer and forward error correction as provided by the MAC layer. V.42bis compression mechanisms, however, remain unchanged.

Data compression attributes are negotiated during the TEI assignment phase of M-ES operation. (Recall these procedures as discussed in Chap. 5, and review Figs. 5-6 and 5-7.) Parameter P0 indicates whether data compression will be enabled or disabled for the duration of the data link connection. If data compression is enabled, P0 also indicates whether the process applies to the forward (only), reverse (only), or both directions of information flow. Figure 6-10 indicates the valid parameter values for V.42bis negotiation. While an M-ES will request compression on both the forward and reverse channels by default, an MD-IS may respond with "no compression," or only one of the channel streams being enabled.

Parameter P1 refers to the number of codewords (known as the N2 variable within V.42bis) supported. Remember that each codeword represents a unique string of characters. CDPD implementations specify a maximum value of 8192 codewords, whereas V.42bis "proper" does not specify a maximum amount applying to N2. For the

V.42bis Negotiation Parameters

Parameter	Default	Minimum	Maximum	Comments
P0	3	na	na	0 = No compression 1 = Reverse direction 2 = Forward direction 3 = Both directions
P1	2048	512	8192	Lowest common value to both M-ES and MD-IS in "powers of 2" accepted.
P2	16	6	250	Lowest common value to both M-ES and MD-IS is accepted.

Figure 6-10   V.42bis variables.

interested reader, Annex II, Sec. II.1 of the V.42bis standard suggests criteria for the selection of the number of codewords a given V.42bis implementation should support.

Parameter P2 identifies the maximum length of a character string represented by an individual codeword (known as the N7 variable within V.42bis). CDPD's default value, as indicated in Fig. 6-10, is a character string length of 16, whereas V.42bis "proper" suggests a default value of 6 characters. Using CDPD default values, M-ES and MD-ISs have the ability to represent as many as 8192 unique character strings, each with a length of up to 16 characters, with a single codeword of 13 bits in length. Note, however, that the set of codewords and their corresponding lengths are set to minimum values upon initialization of the compressing machine. Initial codeword sets and their lengths are set to values of 511 codewords, each 9 bits in length. As a data link connection passes information, the initial dictionary, the number of codewords, and their lengths change dynamically according to the traffic characteristics encountered.

Dictionary space is aligned as a logical set of tree structures. With an 8-bit character format, there will be 256 trees. A tree represents the set of known character strings beginning with a specific character. Each entry (node) within a tree represents one of these character strings. Figure 6-11 is a representation of trees associated with binary strings A, B, BA, BAG, BAR, BAT, BI, BIN, C, D, DE, DO, and DOG. At the highest level of a tree, we see nodes with no "parents." These nodes are called *root* nodes. Entries that are found at the lowest level of a tree are referred to as *leaf* nodes. Each node on a tree has a codeword associated with it that represents the string from the tree's root to the node itself. Thus, Fig. 6-11 provides for 13 unique

V.42bis Tree Structure

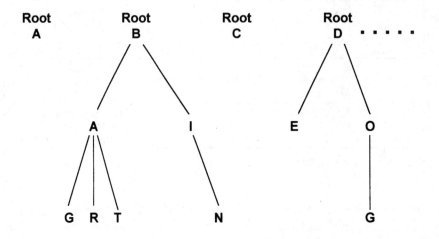

Top (root) row provides codewords for character strings:
A, B, C and D   Total of 4

Middle row provides codewords for character strings:
BA, BI, DE, and DO   Total of 4

Leaf (bottom) row provides strings:
BAG, BAR, BAT, BIN, DOG   Total of 5

This set of dictionary entries provides for the identification of 13 total character strings

Figure 6-11 An example of a V.42bis tree structure.

character strings, each of which may be represented by an individual codeword. Figure 6-12 provides a simplified example of the steps taken to compress the input character string DEN with our sample dictionary entries.

A V.42bis machine must have the ability to leave transparent mode, enter compression mode, and return to transparent mode as character string matches mandate. Octet alignment of uncompressed data is performed by the addition of 0 bits to partial octet fields occupied by codeword substitutes. Individual characters are not compressed; only strings of characters are compressed. This makes sense considering the fact that the default codeword size is 9 bits, assuming a character

V.42bis Compression at Work

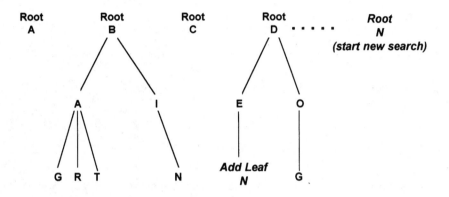

❐ Input characters to be processed: "DEN."

❐ Search for first character, "D," at root. After match is found, progress downwards from root for next character, "E." After match is found, progress down tree looking for match for next character, "N."

❐ Match cannot be found within tree for "DEN." Transmit codeword for "DE" followed by the character "N."

❐ Add the character "N" as a leaf node under "DE."

❐ Assign the next empty dictionary codeword to "DEN."

❐ A new string match search may commence with the character "N."

Figure 6-12 Editing V.42bis trees while performing compression.

length of 8 bits is used. It would be silly to represent an 8-bit character with more than 8 bits. When character string compression is active, a minimum compressible string length of 2 bytes (16 bits) can be represented by 9 bits, for a reduction in overhead of nearly 50 percent. Obviously, as the dictionary grows, so will the length of the trees contained therein. Each time a match is obtained, tremendous reductions in overhead can be achieved via substitution of large sequences of 8-bit characters with small codewords between 9 and 13 bits in length.

The reader should refer to ITU-T V.42bis for a complete description of the data compression mechanism incorporated into CDPD mobile devices.

Registration and Authentication Procedures

Overview

Upon initial channel acquisition, an M-ES must register with the serving network and subsequently be authenticated by the users' home network administration. This chapter focuses on the procedures and protocols involved with the registration and authentication processes. The steps that enable an M-ES to acquire and use CDPD RF channels is discussed in later chapters.

Successful M-ES registration and authentication involves the collaboration of many CDPD processes to provide services on behalf of the mobile user. A *security management entity* (SME) is responsible for the exchange and management of encryption keys, which are vital for protecting the confidentiality of user data. Each MD-IS and M-ES within the CDPD environment supports these SME services. Encryption key management services are identified within the SNDCP header with an NLPI value of 1. SME messages are relayed over the air link with the acknowledged class of information transfer. Using sequenced information frames ensures that data is delivered in the correct sequence, with retransmission support if required.

Mobile-end systems that attempt to register must be authenticated as bona-fide devices via the verification of various static and dynamically changing credentials. Two protocols are used to facilitate M-ES authentication. Between the serving MD-IS and the M-ES itself, a *mobile network registration protocol* (MNRP) provides the necessary services. MNRP is identified within the SNDCP header by an NLPI value of 0.

This M-ES registration protocol also uses acknowledged information transfer at the link level. Between serving MD-ISs and home MD-ISs, the relaying of authentication credentials is accomplished with a *mobile network location protocol* (MNLP). MNLP is also used as a means to manage the mobility aspects of an M-ES. Remember that a user's home function must know where a serving MD-IS is in order to forward packets to that mobile device. MNLP provides the forwarding information required for the home function to reach a mobile device's server. As this protocol is not used over the air link, a subnetwork-dependent convergence protocol identifier is not applicable for MNLP. Instead, the MNLP is supported by the connectionless network protocol ISO 8473. It should be noted that while CLNP is preferred, many first-generation CDPD implementations support only the DOD IP.

Two distinct service sets are provided by the MNLP. One such service is the location update service, which enables a home MD-IS to determine what MD-IS is currently serving one of its "client" M-ESs. A second service, MNLP forwarding, provides a means to encapsulate or tunnel datagrams destined for a given client M-ES within CLNP packets addressed to the MD-IS currently serving that client. Once at the serving MD-IS, user datagrams are "decapsulated," compressed, encrypted, and relayed over a data link connection in the forward direction to their ultimate destinations. Recall that a mobile's serving function and corresponding home MD-IS function may or may not reside within separate physical machines.

Registration and Location Update Procedure

Registering of an M-ES occurs on an NEI-by-NEI basis. The NEI, or *network equipment identifier,* is the network-level address, most probably in the form of a 32-bit IP address in accordance with RFC791. (Refer to App. A.) An individual M-ES may have multiple NEIs associated with it. Multiple NEI capability allows a mobile user to establish sessions with multiple applications, each residing on a unique network. An M-ES may attempt to register after a link level connection between itself and its serving MD-IS has been established via the appropriate SABME/UA exchanges. Exactly "when" a registration attempt will occur will depend on the type of M-ES used and its particular configuration. Depending on its configuration, an M-ES may automatically register NEIs upon channel acquisition or perhaps wait until a user/application request for registration is detected, and then commence the registration process.

Figure 7-1 shows the message format used by an M-ES when registration attempts occur. Registration requests are associated with an individual M-ES by way of the TEI that has been assigned during the

M - ES TEI	INFO	NLPI - 0	ESH, ESB, ISC, ESQ
Address	Control	SNDCP	MNRP PDUs and Options

Format of MNRP PDUs

PDU Type Identifier
Address Length - 7 bytes
OSI Address AFI = 49
IDI = 00
DFI = 00
4 bytes of IP Address
MNRP Options

**MNRP
Message Format**

Mobile Network Registration Protocol

RFC791 IP addresses are presented
to MD - IS within the CLNP address structure.
This applies only to the registration process.

Figure 7-1 The format for mobile system registration messages.

link establishment procedure. As noted above, acknowledged information transfer applies to registration scenarios; accordingly, the control
field is coded with the full complement of command and response
verbs as indicated in Chap. 5, Fig. 5-8. The SNDCP portion of the
message consists of the header that was previously reviewed in Fig.
6-2. Figure 6-3 shows that an NLPI of 0 indicates that these messages
contain mobile network registration protocol dialogue.

Four MNRP messages are used for registration activities. An *end
system hello* (ESH) message is used by the M-ES to request "activation" of an IP address belonging to it. Figure 7-2 illustrates the format of an ESH message. Notice that the IP addresses being registered are presented to the serving MD-IS in a format consistent with
the CLNP protocol, even though the addresses themselves are DOD
RFC791 Internet addresses. OSI addresses may take many forms,
such as binary information, F69 Telex selection digits, X.121 (X.25)
addresses, and IA5 characters. (See App. D.) Within the CDPD environment, OSI addresses have two formats. The first format is used
between ISs and follows the data country code (DCC) syntax (see
App. D). The second format is used between M-ESs and MD-ISs during registration events and follows the local binary syntax. The IDP
portion of the OSI address (also known as an NSAP) indicates local
binary syntax, and the DFI portion of the NSAP indicates the type of
address being identified. Specifically, M-ES registration conveys IP
addresses within the ID portion of the NSAP's DSP. M-ES registration is the only set of activities in which IP addresses are encoded in
this manner. Figure 7-1 illustrates this. During data transfer, IP

MNRP End System Hello - ESH

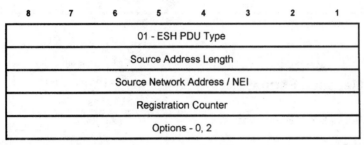

Figure 7-2 The end system hello message is used to request access to the network.

addresses (NEIs) are represented via the IP "proper" protocol, and are subject to the header compression algorithms of RFC1144.

In addition to presenting NEI information, an M-ES that is registering also presents a registration counter and a set of options to the serving MD-IS. (Appendix E contains a description of each MNRP and MNLP PDU, associated options, and applicable result codes.) The registration counter within the ESH message is incremented by a value of 1 every time an M-ES initializes a data link connection after assignment of a TEI. This is referred to as "initial" link level establishment. In the case where a data link connection is simply reinitialized, such as after an FRMR event, an M-ES does not need to request a new TEI assignment. Such a scenario is referred to as *reinitialization*.

Mobile serving functions (MSFs) within serving MD-ISs relay the registration counts found within the ESH message to the M-ES's mobile home function (MHF). Serving MD-ISs relay the registration count to the MHF via the mobile network location protocol's redirect request (RDR) message. Figure 7-3 illustrates the format of MNLP's RDR message. A mobile's home MD-IS checks to ensure that the value is the next expected in sequence number, whereupon further authentication commences.

A *group member identifier option* (GMID) within the ESH is used to correlate individual end systems with a common multicast group. Note that CDPD multicast group NEIs do not mandate the use of an Internet class D address. A CDPD multicast group NEI is an address that has been assigned by the carrier and is used for multicast purposes. A multicast NEI can be any type of IP address assigned by the carrier, including "formal" Internet class D multicast addresses.

MNLP Redirect Request (RDR)

8	7	6	5	4	3	2	1
RDR PDU - Type 1							
Registration Sequence Count							
Source Address Length							
Source Address							
Forwarding Address Length							
Forwarding Address							
Options - 0, 2, 9							

Figure 7-3 Redirect requests are used by a serving MD-IS to request authentication of a user by that end system's home function.

Multicast support in terms of service availability and the format of group addresses are up to the service providers' discretion.

Figure 7-4 illustrates the events that are described below taking place. An authentication parameter option is used to ensure that the M-ES requesting service is a bona-fide device and should be granted access to the network. This option (illustrated in App. E) contains a 16-bit *authentication sequence number* (ASN) and a 64-bit *authentication random number* (ARN). M-ES NEI, ASN, ARN, and GMID information (if a member of a multicast group) are collectively referred to as *credentials*. ARNs and ASNs are preset to a state of 0 upon initial configuration. When registering, these two values are presented to the serving MD-IS and subsequently relayed to the mobile's home function within MNLP's RDR message. A mobile's home function then verifies that the values received are the expected received values for the particular NEI or NEI + GMID in question. If this check passes, the ASN is incremented by 1, and a new ARN is generated by the home MD-IS. Adjusted ASN and ARN values are then relayed back to the registering M-ES via MNLP redirect confirmation (RDC) and MNRP MD-IS hello confirmation (ISC) messages. An M-ES must store the adjusted values noted within the ISC message and place these values in the appropriate fields the next time an ESH is generated. ARN and ASN values are encrypted in both directions when traversing the air interface. It should also be noted that a CDPD service provider will define the frequency of new credential generation, as new ASN and ARN values need not be generated on every registration request. When credential changes do occur, the home MD-IS stores the previous values. This is done in the event

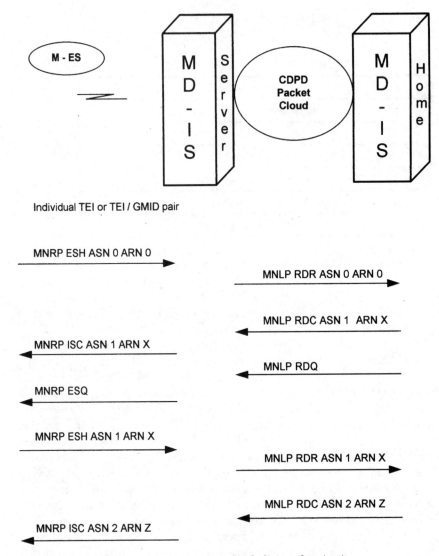

Figure 7-4 An example of registration events, including authentication.

that the ISC message is not received by the M-ES; then the original ESH is retransmitted with the same (nonincremented) values.

Initial registration attempts are handled somewhat differently. When an M-ES generates an ESH with the (default) ASN and ARN values set to 0, the mobile's home MD-IS generates a new ARN and increments the ASN accordingly. After sending an RDC toward the serving MD-IS, an RDQ (redirect query) is generated and sent toward the same server. The purpose of a redirect query message is to confirm that the M-ES is indeed reachable. Upon receipt of the RDQ, the serving MD-IS generates an MNRP ESQ (end system query) toward the registering M-ES. Upon reception of an ESQ, an M-ES generates an end system hello with its current credentials. In this case, the current credentials are the set of values just received as a result of an initial registration request. Eventually, the home MD-IS (or carrier-specific system responsible for authentication) processes this (second) registration request and generates new credentials for the M-ES to use on any subsequent registrations. This two-step process ensures that default credentials are not used by an M-ES for any length of time, enhancing the security of a user's account.

At this time we can see that the MNRP and MNLP work in harmony with each other during registration activities. We also see that the set of MNLP messages enables a mobile's home function to keep track of client M-ES whereabouts. Each time a mobile home function receives an RDR message, the forwarding database is updated to reflect the new location of its client M-ES. Figure 7-5 serves as a cross reference between the MNRP and MNLP interactions that take place during registration and movement of an M-ES between serving areas.

MNRP / MNLP Interaction Cross Reference

PDU	From / To	PDU	From / To	Comments
ESH	M-ES / MSF	RDR	MSF / MHF	MHF checks credentials, issues RDC, updates forwarding table to reflect current server location.
RDC	MHF / MSF	ISC	MSF / M-ES	Contains new ASN/ARN or result code if not accepted.
ESB	M-ES / MSF	RDE	MSF / MHF	Informs MHF that M-ES is no longer reachable via this MSF.
RDQ	MHF / MSF	ESQ	MSF / M-ES	Used to validate reachability of an M-ES or revalidate credentials.
RDF	MHF / MSF	na	na	Used to inform MSF that M-ES has moved to new serving area or holding timer expired

Figure 7-5 Mapping mobile location and mobile registration events.

Deregistration and Table Maintenance Timers

Graceful deregistration of NEIs is the desired way to inform the CDPD network that an NEI is "signing off" from the network and no longer requires service. The process involves an ESB being sent from the M-ES to its current server. Upon reception of the ESB, the server removes any registration entries within its directory databases and sends a redirect expiry message to the NEI's MHF. Entries relating to the current location of the NEI in question are removed from the forwarding location tables. Applications written to work in the CDPD environment may cause an M-ES to deregister during the application's shutdown process. Perhaps a step must be executed by a human operator to cause M-ES deregistration. In any event, it is very possible that an M-ES registers for service but does not gracefully "deregister" by invoking the use of MNRP's end system "bye" message. In this case, a set of timers comes into play that enables the MD-IS sets formally involved with servicing the mobile user to manage the size and accuracy of internal location and forwarding directories.

Within an M-ES is a configuration timer. This timer can be defined by a mobile home function and delivered to an M-ES via RDC and ISC messages. The recommended value for this timer is 4 hours, at the expiration of which an ESH message is sent to the serving MD-IS. Each NEI active within the mobile device causes a separate ESH to be generated. The timer's purpose is to ensure that MD-IS entities are aware of the presence and availability of NEIs even though no data packets have been sent to or from these addresses.

Another timer which is used for location table management is the mobile home functions' holding timer. The holding timer defines the period of time for which location and forwarding directory databases will be maintained for an M-ES that has not been heard from in some time. A value of twice the configuration timer (or more) is recommended. When the holding timer expires, the MHF removes the appropriate NEI from any entries in its local location/forwarding directories. A redirect flush message is then sent to the current serving MD-IS, where registration directory entries for the affected NEI are purged. After these events occur, an MD-IS will be unaware of the location of a given M-ES. Accordingly, M-ESs will need to reregister in order to be serviced after expiration of this management timer.

Interarea mobility management is supported by home and serving MD-IS functions cooperating with each other. Mobile home functions flush out registration information resident in (former) servers while updating internal MHF redirection and forwarding directories.

Intraarea mobility management responsibilities are shared by M-ES and MDBS equipment, and are discussed in the following chapter.

8

Channel Hopping and Management

Overview

This chapter discusses the set of procedures utilized by CDPD components to perform RF channel management. CDPD systems perform two classes of channel management:

1. Acquiring and relinquishing channel space due to the competition between circuit switched and CDPD services.

2. Maintaining continuity of CDPD service when traveling from cell to cell or when a channel being used fails to deliver an acceptable quality of service. Quality of service may be affected by received signal strength indications or other factors affecting the overall bit error rate (BER) performance of the channel.

Our discussion begins with the first "class" of channel management as noted above: allocating resources for competing services. A "view" of channel space allocation within a cell as perceived by both CDPD and circuit switched services is provided in Fig. 8-1. Channels A-1 through A-5 make up the set of operating channels within a cell. From the circuit switched perspective, only channels 2 through 5 are operational, as illustrated in B-2 through B-5. From the CDPD perspective, only channels 1, 2, and 5 are operational, as illustrated in C-1, C-2, and C-5. One of the more interesting aspects of CDPD is that the service's presence should be undetectable by circuit switched voice and data applications. Figure 8-1 shows that while the circuit mode environment is not aware of packet mode services, the CDPD environment does have the ability to detect the presence of other ser-

Topography Knowledge

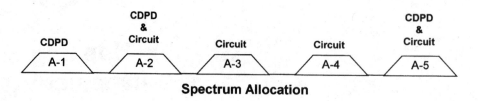

Spectrum Allocation

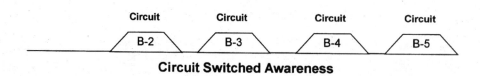

Circuit Switched Awareness

CDPD Awareness

Figure 8-1 CDPD and AMPS viewpoints of cellular channel space.

vices. Obviously, there must be a set of mechanisms to manage the instances when different services are contending for the same channel space.

The first and simplest way to facilitate channel management between the two systems is to dedicate cellular channels exclusively for CDPD use. In this case, AMPSs are not configured to operate on any channel supporting CDPD operation, and packet systems are not configured to operate on channels supporting voice services. This arrangement applies to channels 1, 3, and 4 in Fig. 8-1. The drawback to this approach is that dedicating frequency space for packet services requires many more additional channels—an expensive proposition for the cellular carrier.

At the other extreme, CDPD deployment which is 100 percent overlaid on top of voice channels already in use provides cost savings to the carrier because many of the issues pertaining to cellular topology engineering have already been addressed. Overlaying CDPD services on top of in-use AMPS channel space requires a sharing relationship between AMPS and CDPD services. When channels are shared in this manner, it is the responsibility of CDPD M-ESs and MDBSs to detect a competing service and relinquish channels gracefully. Considering that most AMPSs were in use long before anyone ever thought of a cellular-based mobile IP service, it is easy to see why this responsibility falls onto the shoulders of the M-ES and MDBS components. The positive aspect of this arrangement is that little or no changes need to be made to the existing AMPS infrastructure. The negative aspect to a 100 percent overlaid network is that packet users may experience temporary and intermittent unavailability of service during periods of high voice traffic, such as during "rush hours." Therefore, it seems that a compromise between the two extremes makes sense.

By providing a minimal amount of dedicated channel space, perhaps a single channel per cell for CDPD support, users will always have packet mode access to an RF channel. Any additional channels which are packet mode capable may then share spectrum with circuit switched services, minimizing expenses associated with CDPD deployment. The question before us now is: "How does CDPD effectively share channel space with circuit switched service?" The answer can be found in either one of two scenarios.

The first scenario provides a means for CDPD to determine that a non-CDPD application is using an RF channel. The means of determining such contention can be by passively sampling or "sniffing" a channel in the forward, reverse, or both directions. Such an approach is commonly called *RF sniffing*. The problem with this approach to contention resolution is that it is reactionary: CDPD devices will determine that a conflict exists between services only after a conflict has occurred. There is no foresight to this resolution scheme. Having determined that a conflict exists, CDPD devices must hop rapidly onto another DSMA/CD channel. Failure to relinquish the channel quickly may result in degradation of both packet and circuit mode services. The benefit of this approach is that it is simple, relatively inexpensive to implement, and requires no changes to existing AMPS controllers. If a "sniffer" detects non-CDPD traffic on a DSMA channel, the MDBS controlling that channel immediately ceases transmission. As a consequence, mobile devices using that channel lose synchronization with the DSMA channel stream. Once synchronization has been lost, an M-ES must seek out and acquire a new channel. These events describe what is referred to as a "forced" channel hop. With forced channel hops, no direction is provided to an M-ES by the network, nor is there any forewarning that the DSMA signal will suddenly disappear.

A second scenario for sharing common channel space among CDPD and non-CDPD service calls for an ongoing dialog among AMPS and CDPD controllers. The dialog enables one system to inform the other system of channels which are potential candidates for use, channels which are actually in use, and channels which are next in line to be used if needed. This knowledge can be used to determine channel selection algorithms by both systems and reduce conflicts or "glare" when two services contend for the same channel. A protocol known as the *AMPS channel status protocol* (ACSP) has been designed for this purpose. An obvious advantage of the ACSP approach is that it is not reactionary: At any given time, CDPD and non-CDPD services are aware of the dynamically changing state of RF channel space. Circuit switched and CDPD controllers can plan ahead and solve potential conflicts before they occur. All this means better service for the user.

However, there is a downside to the ACSP approach. Legacy systems would have to be upgraded to accommodate an intelligent interface to the CDPD environment. Many systems in use today may never have such an interface designed for them. For those systems that will be able to accommodate an ACSP interface, expenses will apply. For carriers that have decided to deploy a CDPD system which is manufactured by a vendor other than their existing AMPS supplier, RF sniffing and dedicating CDPD airspace may be the contention resolution of choice. It may be an easier and more affordable proposition to provide ACSP support for carriers that have decided to purchase CDPD controlling platforms from the same vendor that is already supplying that carrier's AMPS hardware. These choices, and the business cases which influence these choices, belong to the carriers. Our examination will be limited to the protocol's message structure and elements of procedure.

AMPS Channel Status Protocol

The channel status protocol used between AMPSs and CDPD systems utilizes a very simple message structure which indicates whether an RF channel is busy or idle. Messages are conveyed between communicating systems via the CLNP using a network service access point (NSAP) identifier of 5. The NSAP value of 5 has been assigned to "channel status services." Figure 8-2 shows the message structure utilized for AMPS-CDPD status messaging. Link level framing mechanisms employed to transfer these messages over individual transmission hops may take many forms. Depending on the physical topologies in use, these messages may be carried by 802.x, frame relay, PPP, or other processes.

AMPS - CDPD Channel Status Message

8	7	6	5	4	3	2	1

01 - Message Type I.D.
CDPD Cell Identifier - 4 bytes
Channel Status 00 = Idle 01 = Busy
Number of channels (255 max) With Channel Status type indicated above
Channel numbers listed in ascending order 1 to 1023 2 bytes for each channel indicated
2-byte Checksum

Carried within CLNP datagram, NSAP = 5. Frame level topology specific.

Figure 8-2 AMPS and CDPD systems may exchange network loading information with each other via the AMPS channel status protocol.

Notification of channel status occurs between CDPD MDBSs and the AMPS. Recall that each MDBS has a link to and is in communication with an MD-IS. Therefore, it is reasonable to allow the MD-IS to serve as an intermediary between the AMPS and CDPD environments, relaying channel status messages to and from CDPD-equipped cell sites. Operation of the protocol is simple. However, it does assume that the AMPS employs a channel hunt algorithm and is aware of an order of selection for channels to be placed in use. MDBSs and the cooperating AMPS simply notify each other each time a channel is placed into use and/or made idle via the ACSP channel status message. In addition to notifying MDBSs with busy-to-idle state changes of channels, the AMPS also periodically provides an MDBS with a summary of each idle channel known at the time. This is done by incorporation of a channel status timer which is configurable by the service provider. When providing an MDBS with a summary of idle channels, an AMPS should list the channels in the anticipated order of assignment for future use. Doing so will assist an MDBS in deciding what channels can be considered "best" channels to use in the event a migration of M-ES units is required. Figure 8-3 illustrates a series of handshakes between MDBS systems and AMPSs which serve to notify each other of channel state changes.

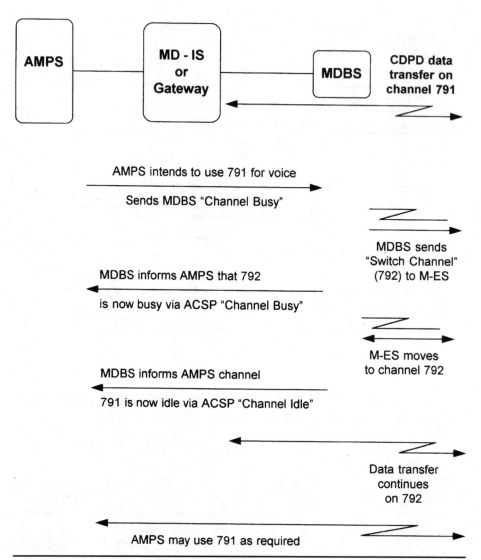

Figure 8-3 An example of AMPS-CDPD handshakes that each system can use to inform the other of changes in RF channel use.

M-ES and MDBS Channel Management

As noted above, the second class of RF channel management consists of a set of procedures that ensure continuity of service when traveling from cell to cell or when CDPD quality is degraded. Unlike the AMPS, control and administration between MDBS and M-ES devices occur inband. Messages that are transmitted which have to do with these operations and management (OAM) activities are distinguished from user data via service access points reserved for these purposes. Information that an M-ES requires in order to select the best CDPD channel over which to operate is periodically transmitted over each DSMA/CD channel. This includes topology information which enables an M-ES to determine alternate channels to hop to when the current channel's signal quality is degraded. These periodic broadcasts enable M-ES equipment to move from channel to channel without any assistance or directives from the CDPD network itself. Periodic broadcast of topology information allows M-ESs to make intelligent and independent channel movement decisions. Of course, it is unwise to rely solely on the capabilities of user equipment to protect cellular air space. Accordingly, CDPD must also provide a means to forcibly direct both individual and groups of mobile devices from one channel to another. Together, both the network and mobile devices cooperate to ensure availability for both packet and circuit switched service.

The remainder of this chapter focuses on how the network and mobile devices collaborate with each other to maintain acceptable levels of service. Our discussion continues by examining the message set between the network itself and the M-ES population. Once we understand the dialog that takes place, we will be able to appreciate the decision-making capabilities of CDPD equipment.

During the time that an M-ES is synchronized to a CDPD channel stream, mobile devices build a database containing topology information about the current serving cell and that cell's adjacent neighbors. As mentioned earlier, this information is used by the M-ES when a move to a new channel is required. However, how does an M-ES initially acquire a CDPD channel stream? Search algorithms for initial channel acquisition have not been standardized. An M-ES may search all available channels in a sequential manner, search only the A or B carrier frequencies, or perhaps refer to a list of channels that has been predefined. Once initially synchronized to a CDPD channel, convergence time of subsequent M-ES power-ups may be minimized by scanning channels that are known to be CDPD capable. For example, such information may be obtained in the topology database which is (hopefully) maintained by an M-ES when using the service. Topologies change, of course; such is the nature of the cellular environment. Also, a user may be in a different serving area when attempting to use an M-ES, rendering historic topography information useless. However, in the event that a user more often than not registers for service in the same general area, historic topography information may indeed assist in minimizing convergence time, especially when dedicated channels are available. As vendors are free to implement their own algorithms, differences in implementation can be expected. When comparing the ability of two M-ESs to find, synchronize, and register on a channel, the differences in convergence times may not be significant or even apparent to application-level services. In a telemetry system, mobile-end systems can be expected to remain constantly powered up and, in many cases, stationary. In this scenario, the ability to perform quick, reliable channel hops becomes more important than rapid initial channel acquisition. For mobile professionals and other users of laptop PCs or similar devices, initial channel acquisition will most likely occur well before the machine completes its "boot" processes and the application is ready for user input. Again, it is easy to see that in most cases, the channel hop capabilities of an in-use system will have greater impact on users' perceived performance as opposed to initial channel acquisition.

Wide Area Channel Search

After initially acquiring a CDPD channel stream, mobile devices perform what is called a *wide area channel search*. Wide area channel searches are performed when there is no configuration (topology) database to use which points to a known CDPD channel. This search is also used when a database contains information (values) that are inconsistent with the serving area. An example of this scenario is when a mobile device which is "homed" out of Carrier A on the East Coast of the United States is powered up within a different carrier's serving area on the West Coast of the United States. Any historical topology information which may have been saved during the last East Coast session will certainly not be valid when attempting to use West Coast resources. Service provider network identifiers, cell identifiers, and other parameters which have been stored will not match those detected within (new) current OAM messages. Hence, the database is of no assistance at this time; procedures consistent with initial channel acquisition now apply, and the M-ES executes its channel scan algorithm. Simply put, the wide area channel search begins by looking for a channel with acceptable RSSI (receive signal strength indication; minimum, 113 decibels below 1 watt). Once a signal has been found, an attempt to synchronize to DSMA channels is performed. If that is unsuccessful, the search continues until a positive result is found. Once it is successful, the M-ES confirms that the block error rate (BLER; nominal default, 20 percent) of the channel is acceptable. Once successfully synchronized to a channel that meets the BLER threshold criteria, mobile devices wait for a channel stream identification message.

Figure 8-4 identifies the contents of the channel stream identification message. These messages are broadcast periodically by the MDLP layer via UI frames/TEI = 0 and do not require the use of a network-layer routing protocol. These messages are broadcast on the forward channel stream approximately every 5 seconds and serve as a means for mobile devices to determine "where" they are operating. In

Channel Stream Identification Message

(Sent in MDLP Layer 2 / TEI = 0 Broadcasts)

8	7	6	5	4	3	2	1
LMEI = 42 (2A hex)							
Message Type = 0							
Protocol Version = 1							
Dedicated 1 - yes 0 - no	Capacity 0 = Vacant 1 = Full	Channel Stream Identifier (CSI)					
Cell Identifier SPNI + Cell number							
Service Provider Identifier - SPI							
Wide Area Service Identifier							
Power Product							
Maximum Power Level							

Figure 8-4 After initial acquisition of an RF channel, an M-ES must observe a channel ID message before frames can be transmitted.

order to operate on a CDPD system, an M-ES obviously must be enabled by an authorized party. Some of the issues involved in enabling an M-ES include assignment of NEIs (IP addresses), GMIDs for multicast purposes, and resetting validation credentials as per earlier discussion. In addition, an M-ES may be programmed with variables that assist in customizing the operating environment of the device. Examples include programming allowable network service providers to register with (SPI), regions within which such service providers have domains (SPNI), and A/B carrier preferences. When a mobile device receives a channel identification broadcast, the parameters in the message are compared to any profiled information that may exist within the M-ES itself. If there are conflicts among the allowable set of parameters an M-ES has been programmed to operate with and the corresponding information received in the channel identification message, use of that CDPD channel is prohibited.

An additional parameter included in the channel identification message is a flag to indicate whether the channel is a dedicated CDPD channel or a channel which is shared with circuit switched service. It is possible that a manufacturer of CDPD mobile-end systems can use this information to give an end user the ability to define a preference for a particular channel type. Also, network service providers would determine capacity thresholds for each channel

based on geography and population criteria. Within the channel identification message there is a capacity flag which indicates to an M-ES
if a channel can accommodate additional users, or if the load factor is
too high. The specific CDPD channel stream being referred to is indicated in the 6-bit field immediately following the capacity flag.

Next, there are three related fields in the channel identification message. The cell identifier is a combination of a 16-bit *service provider network identifier* (SPNI) which is globally unique (see App. A) and a 16-
bit *cell number* which is assigned by the service provider itself. This
information identifies the specific cell within a specific carrier's serving
area that is providing the DSMA channel. A *service provider identifier*
(SPI) is assigned for each carrier. An SPNI represents one of the many
individual CDPD networks managed under the authority of a particular licensed carrier of CDPD service. In many instances, geographic
coverage requires the collaboration of multiple carriers acting together
to provide service for the population of that area. Also, when such
arrangements exist, it is possible that special billing arrangements will
apply. The *wide area service identifier* (WASI) is used as a marketing
identifier to indicate if such an arrangement exists.

M-ES Power Determination

Last, there are two fields related to RF power, the power product and
maximum power level fields. This information is provided as a means
to enable an M-ES to determine what its output power should be set
to just before any transmission burst. M-ESs should not adjust power
in the middle of a burst on the reverse channel. Instead, a calculation
should be made to determine the appropriate output power immediately before each burst. First, recall that each M-ES belongs to a category of mobile equipment classes. The acceptable maximum output
power for each class of mobile device is indicated in Fig. 8-5.
"Maximum power level," as indicated in channel identification messages, informs M-ESs of the highest tolerable power levels permitted
(0 through 10), regardless of the equipment class to which the M-ES
belongs. Computing the power levels which are actually used, however, requires some arithmetic. An MDBS observes the mean levels of
all signals received from M-ESs on the reverse channel. The mean
signal strength derived on the reverse channel is subtracted from
-143 dBW (decibels below 1 watt). This difference becomes the
"power product" that is contained within channel identification messages. Once an M-ES calculates the difference between -143 dBW
and the value supplied in the channel identification message's power
product field, the M-ES subtracts its own received signal strength
indication of the forward channel from this value. The result of this

M - ES Nominal ERP Levels by Device Class
Values in terms of [dBW]

Power Level	Class I	Class II	Class III	Class IV
0	6	2	-2	-2
1	2	2	-2	-2
2	-2	-2	-2	-2
3	-6	-6	-6	-6
4	-10	-10	-10	-10
5	-14	-14	-14	-14
6	-18	-18	-18	-18
7	-22	-22	-22	-22
8	-22	-22	-22	-26 +/- 3
9	-22	-22	-22	-30 +/- 6
10	-22	-22	-22	-34 +/- 9

Figure 8-5   Maximum M-ES output power ranges by equipment class.

computation is the correct output power for an M-ES to use for reverse channel transmission bursts. An example is offered:

- MDBS's mean RSSI = −85 dBW.
- Difference between −143 dBW and −85 dBW = 58.
- Encode the value 58 in binary in channel ID message's power product field. This represents the "delta" between 143 as a reference and MDBS's received power.
- M-ES observes received signal strength at −80 dBW.
- M-ES subtracts 80 from 85 and determines that the proper output power for transmission is −5 dBW.
- M-ES "rounds off" −5 dBW to the nearest legitimate power level, which is 3. M-ES will output at −6 dBW.
- M-ES will restrict itself to a maximum power of channel identification message's "maximum power level" field if above arithmetic results in a stronger value.

Completion of Channel Acquisition

After the events described above have taken place, an M-ES determines if it is operating within the same cell and serving area that its current topography information database reflects. Assuming that there is no difference between the topography information stored in the M-ES's database and the information supplied in the channel identification message, normal operations continue. If the M-ES realizes that it is operating within the same serving area but the cell itself has changed, the topography database is initialized and updated appropriately. If an M-ES determines that it is currently within a new serving area, the mobile device must register with the new server. Of course, mobile network registration will be attempted only if the M-ES has determined that it is operating within an allowable SPI/SPNI set.

M-ES Topology Database Information

We have discussed the procedures taken by an M-ES when initially acquiring a channel and attempting to use the CDPD service. We have also seen repeated references to a "topology database" which contains vital information pertaining to the network environment currently serving the user. We will now discuss how this database is built and the contents contained therein.

Recall that topology information is broadcast periodically over each DSMA channel stream. The information itself is conveyed with a "cell configuration" message. The purpose of the cell configuration message is to provide M-ES devices with enough information about the cell in which they are operating (as well as adjacent cells) that they can make intelligent, unassisted decisions when a hop to another channel is required. Figure 8-6 illustrates the format of the cell configuration message. The messages are broadcast over each CDPD channel stream at intervals defined by the operating carrier. These messages describe the topologies of an M-ES's current cell and that cell's adjacent neighbors. Figure 8-7 gives a visual idea of the scope of knowledge an M-ES may learn from these messages. Examination of the cell configuration message tells us that there is a lot of information to be learned by an M-ES about its RF environment. We will begin by discussing the parameters contained in this message on which we have not yet elaborated.

A cell configuration message identifies the area color code (MD-IS) supporting the cell as well as the total number of CDPD DSMA channels that the cell supports. The "face" bit identifies whether an adjacent cell is considered to be a "face neighbor" or an "adjacent neighbor." Face neighbors share a common antenna location, while adjacent neighbors utilize different cell sites. A face neighbor indicates the use of sectored, or directional radiation as opposed to omnidirectional radiation. The configuration lists being distributed include all of the channels that are CDPD capable, but not necessarily supporting CDPD service at that moment. Channels which are dedicated for CDPD use are identified as such via a dedication bit which accompanies each listed channel number. Note that there is a field within

Cell Configuration Message

(Sent in MDLP Broadcasts / TEI = 0)

8	7	6	5	4	3	2	1
LMEI = 42 (2A hex)							
Message Type = 01							
Cell Identifier (SPNI + Cell Number)							
Face	0	Active Channel Streams			Area Color Code		
Reference Channel							
ERP Delta							
RSSI Bias							
Power Product							
Maximum Power Level							
Dedi-cated	Reserved						
First Listed RF Channel Number							
Intermediate Channel Numbers in List							
Dedi-cated	Reserved						
Last RF Channel Number in List							

Figure 8-6 Cell configuration messages convey topology information about neighbor cells to in-use M-ES equipment.

the cell configuration message indicating the number of channels which are CDPD active. The maximum value that can be carried in this field is 7. However, there may be more than 7 CDPD channels which are CDPD capable within a given cell. Therefore, a value of 7 in the active channel stream field may indicate that "more than" 7 CDPD channel streams are available.

Originally, a CDPD mobile-end system would hop to another channel only if the current channel is deemed to be "unacceptable." Reasons to consider a channel unacceptable include subpar BER performance or perhaps conflicting color code information being detected. This posed an interesting dilemma in that there were no mechanisms in place to ensure that at any given time a mobile device was making use of the "best" channel out of the many channels which might be available. Once synchronized to a channel stream, the M-ES would maintain operation over that channel stream until it was no longer possible. Meanwhile, there might have been another channel in another cell which would have provided a stronger signal. A set of mechanisms is now defined

Cell Topography Information
Learned by an M-ES operating in cell 20 - 6 - 13

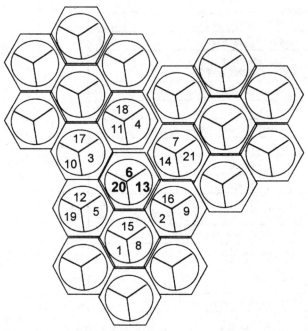

Figure 8-7 An example of topology information which can be derived from cell configuration messages for a current cell (center) and its neighbors.

which will assist an M-ES in acquiring the most appropriate channel at any given time. The act of determining which channel is the "best" channel to use is called an *adjacent channel scan*. Inclusion of a reference channel within cell configuration broadcasts is one of the mechanisms that assist an M-ES in making such an intelligent decision.

During normal operation, an M-ES periodically scans reference channels as indicated in cell configuration messages for signal strength measurements. An M-ES will relocate to a different channel if it determines that another cell can provide better service. Two types of reference channels may apply. If a cell supports a dedicated CDPD DSMA channel, that dedicated channel can be used as the reference channel, as it is being modulated continuously with no interruptions. However, if a cell does not employ dedicated CDPD channels, finding a DSMA channel stream to measure may be troublesome. In an environment where CDPD is shared with circuit switched service, DSMA channel streams will appear and disappear rapidly because of channel changes. Given the nature of CDPD channel hopping, it is permissible to use a known, fixed channel such as an AMPS control channel for signal strength (reference) measurements. Because the reference channel may not be a CDPD channel, there is an *effective radiated power* (ERP) delta field within cell configuration messages. This field informs M-ES devices of the difference between power transmitted on the reference channel and power transmitted over a CDPD channel, which may be different. M-ESs are expected to take into account the difference between the two values when assessing the power of a DSMA channel stream.

An additional parameter which should be taken into account is the RSSI bias field. The concept of a bias value enables a service provider to define geographically where a channel hop boundary between adjacent cells will occur. Because of population density and terrain features, it may be desirable to influence a mobile's decision on which cell to change to using a measurement other than pure signal strength. An example of this is a scenario where an M-ES needs to acquire a new channel because of poor quality. The M-ES has a choice of two (new) signals of relatively equal strength, each in separate adjacent cells. It is possible that one cell services a high population area while the second cell services a lower density of users, perhaps due to the presence of water, forest, etc. With a bias adjustment value, mobile devices can be persuaded to choose the less populated cell over a cell which could be expected to experience higher congestion levels. Note that while ERP delta is a measurement between CDPD and reference channels, the bias information reflects an adjustment to comparisons made between CDPD channels, and is defined by the cellular operator in decibels above 1 watt. In this manner, the operator can control channel change boundaries. Figure 8-8 shows the effect of using biasing information to influence channel change decisions.

Cell configuration message "power product" and "maximum power level" fields provide information similar to that seen in channel identification messages. The difference is that these values are for adjacent cells as indicated in the cell identifier field.

Maintaining Current Channel Continuity

Of course, moving to a more suitable serving cell when one has been detected makes good sense. When such a move is called for, we have seen how a CDPD mobile device uses topology information to make intelligent decisions on new channel acquisition. However, interchannel movement should be controlled so that mobile devices are not constantly moving from cell to cell (and back again) because of relatively small changes in signal strength. Another enhancement to CDPD is the inclusion of mechanisms that help an M-ES determine not only "where"

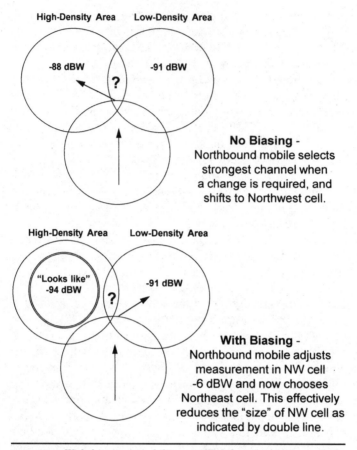

High-Density Area   Low-Density Area

-88 dBW   -91 dBW

?

No Biasing -
Northbound mobile selects
strongest channel when
a change is required, and
shifts to Northwest cell.

High-Density Area   Low-Density Area

"Looks like"
-94 dBW   -91 dBW

?

With Biasing -
Northbound mobile adjusts
measurement in NW cell
-6 dBW and now chooses
Northeast cell. This effectively
reduces the "size" of NW cell as
indicated by double line.

Figure 8-8 With biasing capability, a cell's "shape" or coverage area
can be tailored.

a better serving cell is, but "when" it makes sense to move over to that
cell. Periodic broadcasts of channel quality parameters helps a mobile
device to ascertain acceptable performance attributes of the current
channel stream. When an M-ES determines that the indicated performance attributes can no longer be met, a search for new channel space
begins. The channel quality parameter message is indicated in Fig. 8-9,
together with suggested default values. One of the parameters included
in this message, RSSI hysteresis, is used by an M-ES to lean toward its
current serving channel or a new (neighbor) channel when making signal strength measurements. This value tells an M-ES how much difference in signal strength is required between two channels before a
change should take place. Thus, we can see how it is possible to minimize the number of cell-to-cell transfers that take place.

RSSI scan time informs mobile devices what the maximum allowable time frame is between M-ES assessments of its received signal

Channel Quality Parameters Message

(Sent in MDLP Layer 2 / TEI = 0 Broadcasts)

8	7	6	5	4	3	2	1
LMEI = 42 (2A hex)							
Message Type = 02							
RSSI Hysteresis							
RSSI Scan Time							
RSSI Scan Delta							
RSSI Average Time							
BLER Threshold							
BLER Average Time							

Configurable RF Resource Parameters

Parameter	Default
RSSI bias	0 dB
Power product	35 dB
Maximum reverse channel power	0
RSSI hysteresis	8 dB
RSSI scan time	90 seconds
RSSI scan delta	8 dB
RSSI average time	5 seconds
BLER threshold	20 %
BLER average time	5 seconds

Figure 8-9 Channel quality broadcasts assist an M-ES in choosing a "best" channel for packet data transmission.

strength. RSSI scan delta informs an M-ES when to assess signal levels of adjacent channels using stored topology information. It is important to make frequent checks of signal quality to ensure that an M-ES is always synchronized to the best channel available. However, it is unwise to burden an M-ES with constant signal assessment processing at the expense of data transfer capability. So, a mobile device measures its received signal periodically. Let's say that this occurs at the default 90-second time interval. Then, when a difference of [RSSI scan delta decibels] occurs, it checks to see if this difference is maintained for a duration in excess of [RSSI average time]. If the noted difference in received power level exceeds this average time frame, then an adjacent channel power scan is initiated. At this point, the mobile device is assessing the best new cell to hop to in the event the current operating environment does not improve. A move to an adjacent cell will occur once assessing the best cell to hop to has been completed *and* the [RSSI hysteresis] value has been exceeded.

In a similar fashion, signal quality parameters [BLER threshold] and [BLER average] are used to determine when a hop should occur due to data integrity issues as opposed to signal strength issues.

DSMA/CD Channel Attributes

Recall our discussion of the DSMA/CD channel access mechanism as well as the discussion of MDLP at CDPD's link level. Both of these processes retransmit if a prior transmission was deemed to be unsuccessful. The DSMA/CD access mechanism waits for an "idle" condition to be sensed before a reverse channel transmission burst begins. Confirmation that the burst was successful is indicated in a decode flag. If the decode flag indicates "unsuccessful," the MAC layer waits for another idle condition and tries again. Remember that multiple retries may occur, with a delay between each attempt due to idle "wait" conditions. Meanwhile, while all this is going on at the MAC level, MDLP at layer 2 is waiting for a response to frames it has transmitted. If the MAC layer cannot complete its burst within a certain number of retries, MDLP eventually begins retransmitting. Duplicate frames are then handed to DSMA/CD, starting the cycle all over again.

How does this relate to the current discussion of RF channel management? The DSMA/CD access mechanism is based on a wait/try/try again repertoire. CDPD mobile devices need to know "how long do I wait for an idle?" They need to know "how many times do I reattempt to gain access to a channel?" Time-outs, retry counters, burst sizes, and the like all need to be understood. Accordingly, the channel access parameters message exists. This message allows an M-ES to discover channel-specific DSMA/CD operating guidelines. The message structure is illustrated in Fig. 8-10. Periodic broadcasts of these parameters (every 60 seconds by default) are sent in the forward direction. Information conveyed in these messages allows physical medium access mechanisms to determine when they should defer to link level for transmission recovery activities.

Maximum transmission attempts defines how many times an M-ES may seek an idle status flag. If this value is exceeded, the link level must be able to recover. Such a recovery may be in the form of a time-out and retransmission of frames pending acknowledgment. *Maximum blocks* limits an M-ES to a total transmission burst length. Any blocks sent in excess of this value will result in a decode unsuc-

Channel Access Parameters Message

(Sent in MDLP Layer 2 / TEI = 0 Broadcasts)

8	7	6	5	4	3	2	1
LMEI = 42 (2A hex)							
Message Type = 05							
Maximum Transmission Attempts (13)							
Minimum Idle Time (0)							
Maximum Blocks (64)							
Maximum Entrance Delay (35 microslots)							
Minimum Count (4)							
Maximum Count (8)							

Figure 8-10   CDPD media access parameters with (default values).

cessful flag being sent, which in turn results in a retransmission attempt. *Maximum entrance delay* defines the upper delay limit in "microslots." A microslot is the 60-bit time duration between the last bit of busy/idle status flags. At 19.2 kilobytes per second, this is 3.125 milliseconds. Recall that an M-ES will wait a random number of microslots before attempting access to a channel, or sensing the busy/idle status flag. The entrance delay parameter sets the range of this random number used for initial channel access. *Minimum and maximum count* fields in the channel access message define the exponent in the retransmission mechanism employed by DSMA. An M-ES will "exponentially" attempt access to CDPD channel streams. Specifically, the minimum count defines the smallest range of random number generation, while the maximum count defines the largest range of random number generation. The range of random number possibilities is defined as 2 to the [count] power. With each unsuccessful retransmission attempt, the [count] exponent will be incremented until the maximum count limit has been reached. Numbers in parentheses in Fig. 8-10 identify the CDPD-specified default values for these parameters.

Obviously, when an M-ES hops to another cell, the new physical channel access attributes preferred by an MDBS may be different from those being used before the hop occurred. With this in mind, an M-ES may continue to operate under the "old" set of guidelines until it has learned via channel access parameter broadcasts any new set of guidelines that may apply.

Switch Channels Message

(Sent in MDLP Layer 2 / TEI = 0 Broadcasts)

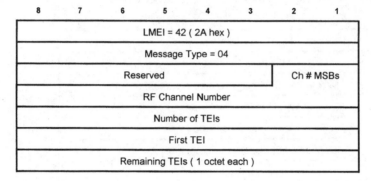

8	7	6	5	4	3	2	1
LMEI = 42 (2A hex)							
Message Type = 04							
Reserved						Ch # MSBs	
RF Channel Number							
Number of TEIs							
First TEI							
Remaining TEIs (1 octet each)							

Figure 8-11 An MDBS can force an M-ES to hop to another RF channel via the *switch channels* message.

Directed Channel Hops

So far, we have seen how a CDPD mobile device is able to gather information related to its operating environment. We have seen how an M-ES uses this information to intelligently determine when an unassisted hop should occur. Finally, we have seen the considerations involved which enable an M-ES to determine where the best, most appropriate channel streams to use reside. Not all channel hops are unassisted, however. As mentioned in our discussion of the ACSP, the CDPD network does have the ability to direct M-ESs to new channels when the need arises. We conclude this chapter by discussing the procedures involved in MDBS directed channel movement.

Many factors can cause an MDBS to determine that a channel hop is required. Some of these reasons include contention among CDPD and circuit switched services and network-specific capacity allocation plans. For instance, it is possible that a service provider utilizes directed channel hops for load balancing or to manage time-of-day (such as rush-hour) requirements. Regardless of the reason for invoking such procedures, MDBS-directed channel hops utilize the *switch channels* message provided by radio resource management services. The message enables a CDPD operator to direct one or a population of mobile devices to new channels. The message, very simply, states the channel to move to, and the TEIs that should move to the indicated channel. If every TEI being served at the moment is being directed to switch channels, the TEI indicator is set to indicate "broadcast," or 0. Figure 8-11 provides a template for the switch channels message.

9

Applications

CDPD Users

It is commonly said that CDPD applications fit into one of two major categories. The first category is the commercial or industrial category. The second category is the mobile professional category.

Let us first consider the mobile professional. Mobile professional application sets consist of two major components. The first component within a mobile executive's application suite is some form of messaging or mail service. The second component consists of software clients that are compatible with users' corporate network operating systems, interacting with in-place host-resident applications and databases. Terminal emulation, remote node, and remote control of PC platforms are possible ways to provide this type of connectivity. In many cases, mobile communications with these processes may be better served with a circuit switched solution, simply because of the volumes of data being generated. This is especially true if flat-rate pricing is not available for CDPD. Usage-based pricing can present some problems. Obviously, metered network use is unattractive if applications require large volumes of data to be transferred. Second, it is difficult to design a mobile network solution when volumes of data cannot be controlled or predicted. As a result, mobile professional communications can be expected to take a slower migration path to CDPD service.

Two industry changes will have to occur before CDPD fits the mobile professional profile as comfortably as circuit switched solutions do today. Applications on portable PC platforms will require more streamlined, "network-efficient" versions introduced. In fact, this is one of the big opportunities for today's software developers. Wireless communications is definitely gathering momentum. If wireless is here to stay, a huge amount of revenue will be generated by simply taking existing applications and introducing network-efficient (mobile) versions of the same products. This is already happening. A second change that needs to occur is large-scale deployment of CDPD network availability. The availability of CDPD must also be coupled with interworking functions among carriers. By definition, the mobile professional is a frequent traveler. Without coverage and internetworking capabilities, CDPD will not be embraced by the mobile executive as has circuit switched cellular.

Considering the above, it is easy to see why many early CDPD solutions fit the industrial or commercial profile. For many commercial applications, "roaming" capability may be limited to smaller serving geographic areas. Point-of-sale (POS) terminals, hand held scanners, and other highly portable personal digital assistants (PDAs) will be widely used for these applications. Unlike the mobile professional, a taxi dispatch operation in Phoenix, Arizona, will probably limit its operation to the greater Phoenix area, and a power utility in New Jersey most likely will not require national coverage (at least not under current rules and regulations). This aspect of industrial applications minimizes the requirements for ubiquitous coverage and interoperability between carriers. Also, many applications that fit the commercial or industrial profile make use of short transactions with a minimum amount of time per "session." As discussed in Chap. 2, these attributes create a good fit with packet-mode network solutions. Finally, many commercial applications, point of sale (POS) and telemetry (remote monitoring), for example, can be put to work by numerous organizations, regardless of their line of business, as they tend to be similar in nature. For instance, what works for the taxi operation in Phoenix will probably also work in other cities. What works for a power utility in New Jersey may also work for Niagara Mohawk in New York State, and so on.

Mobile-End-System Types

Just as there are two classifications of CDPD users, there are two classifications of CDPD mobile-end systems. The mobile professionals' M-ES will be "briefcase compatible," whereas commercial M-ESs will not necessarily be so. The term "briefcase compatible" is used to describe compact assemblies which are easy to carry about. Three generic types of "professional" M-ESs are available: integrated, external, and PCMCIA.

An integrated M-ES is installed within a portable computer or other laptop or palmtop device, and hidden from the users' view. For example, in a PC environment the M-ES itself may be built to occupy a bay intended for a hard disk drive, with the antenna built into the frame of an LCD display. Power would most likely come from the host platform itself, making battery life an issue because power would need to be shared among M-ES and host functions.

PCMCIA adapters, simply because of their size, will most certainly require external antenna assemblies. These external assemblies would also accommodate batteries to power the entire CDPD system, helping to preserve host power sources. DC power converters will interface with the external assemblies to facilitate charging and battery conservation. When choosing either an internal or PCMCIA CDPD (or circuit switched) modem, it is important to verify that the host computer is "wireless friendly," and will not be adversely affected when operating in close proximity to a cellular radio. One way to verify this is to ask your M-ES vendor if any particular systems have been tested and certified with the M-ES you intend to use.

External M-ES equipment will consist of a small device similar in size to a portable telephone. As in cellular telephones, antenna assemblies will be retractable. Connectivity to an externally attached host will be via serial communications cables such as the industry standard RS232 interface. Power will be supplied via the M-ES's own DC battery source and/or via AC adapters. These devices may be clipped temporarily onto a laptop computer, or simply rest next to the host on a table top.

M-ES equipment should provide a means for the user to quickly and easily determine signal characteristics and operation of the device. LED (light-emitting diode) or similar displays should enable the user to determine "at a glance" two important pieces of information. First, the user needs to be able to determine if a CDPD signal has been acquired. Second, the user should have an indication of whether the acquired signal is poor, marginal, adequate, or good. All M-ESs will allow users to gain access to signal acquisition and power information via use of AT (attention)–type command sets. However, the mobile professional does not want to be bothered with sending special commands to an M-ES or closing and opening up "windows" to view this information. Remember that a user should not have to think about the communications medium that he or she is using.

Providing external indication such as an LED display for PCMCIA and internally mounted M-ESs is a bit more difficult. In the case of PCMCIA cards, LED displays will have to appear on an external antenna assembly. Processing of the CDPD signal will occur on the card itself. This introduces a need to build a hardware control channel in the connection between the host and antenna assemblies, which would control the externally visible LEDs. With internally mounted devices, there would be no way to view any hardware component. Accordingly, a viewable status bar or "window" would be the best alternative in this scenario. Of the two choices, a status bar that is permanently visible makes more sense. An old truism is called to mind: "Users" want to "use" technology, "users" should not be expected to "manage" technology, and "user friendliness" increases as the knowledge required to implement a technology decreases.

Consider for a moment that mobile professionals may resist having to learn proper use of new technologies to do their job. From this perspective, it makes sense to design a new product to have the look and feel of an existing, well-known product. To the extent possible, a CDPD interface should look much like a telephone. Some examples of this thought are presented. For example, use circuit switched audio tones that are familiar to the user to provide feedback. When registering, for instance, the sound of dial tone and digits being dialed "means something" to a user. If a registration fails, the sound of a busy tone "means something" to a user. If feedback can be given to users of new technologies that mimic the behavior of accepted technologies, user acceptance will come more quickly. This is especially true when an end user has purchased a multimode modem which supports wireline, circuit switched wireless, and CDPD operation. Users should not be expected to become familiar with multiple personalities of "modems," even if multiple network technologies are utilized.

Mobile-end systems intended for industrial or commercial use may not require the cosmetic appearance that is desirable for mobile professional use. In many cases, this equipment may be permanently mounted within a vehicle, building, or shed. Power for an M-ES of this type may be derived from existing power sources in the vehicle or building itself. Antenna assemblies may be of the "car phone" variety; they need not be "briefcase compatible." Applications that make use of these M-ESs will most likely be fixed in the sense that they are written to perform a limited set of specific tasks. Examples include credit card transactions, remote meter reading, and vehicle dispatch. Direct human interaction with the M-ES will therefore probably be limited. Accordingly, LED diagnostic information may not be as comprehensive or as important as in the mobile professional's M-ES, but it should be obtainable via AT commands. Because many of these devices will not be attended by humans, the device itself and/or applications connected to the devices should be manageable via remote processes such as SNMP.

Duplex versus Half-Duplex Operation

M-ES equipment may operate in the full- or half-duplex modes. A full-duplex device has the ability to transmit and receive simultaneously. A half-duplex M-ES operates in a Ping Pong mode; it can only transmit or receive at any given time. Understanding the behavioral characteristics of an application is important when choosing an M-ES type. For instance, many industrial applications operate in a Ping Pong–type manner. Credit card authentications and remote monitoring are two examples of a pure "query/response" environment. For these types of applications, a half-duplex M-ES will provide the same level of performance as a (perhaps) more expensive full-duplex M-ES.

Mobile professionals, however, will probably benefit most from a full-duplex-capable M-ES. This is true because a human being can be expected to "browse" for information within the various databases they can establish sessions with. Also, many existing corporate applications make use of TCP/IP, or other windowing, full-duplex protocols. Use of a half-duplex M-ES with these types of applications may contribute to decreased throughputs and increased response times, because of the possibility that, while receiving data segments, TCP or some other process resident with an F-ES may attempt to send toward a mobile device an acknowledgment or other traffic. If these messages happen to be transmitted over the air interface as a half-duplex M-ES is still using a reverse channel, time-outs and retransmissions will result. In such a scenario, file transfers in the mobile- to fixed-end system direction would most likely suffer greater impact than file retrievals in the F-ES-to-M-ES direction.

Users should consider the behavioral characteristics of their applications and associated protocols that will be transported via CDPD when making M-ES (or any network component) purchasing decisions.

Physical Connections

A CDPD mobile-end system consists of three main components. A *mobile application subsystem* (MAS) is made up of application processes which are external to the CDPD-specific protocols. The MAS may be a laptop computer connected to a CDPD "modem" via RS232 serial cable. Within the laptop (or other MAS) may be multiple *subscriber applications* (SAs). MAS processes may be external to the CDPD such as in the above laptop example. An MAS may also be integrated into a mobile device such as in the case of a POS terminal.

In either case, an MAS will need to interface with a *subscriber unit* (SU). The subscriber unit contains the CDPD-specific processes and protocols that we have been discussing up to this point. An SU always resides within the CDPD mobile "device." In our laptop example, the SU is what communicates with the laptop via the serial RS232 cable.

Also found within the CDPD device is a *subscriber identity module,* or SIM. The SIM contains information specific to that mobile device, such as authentication information, NEIs, authentication credentials,

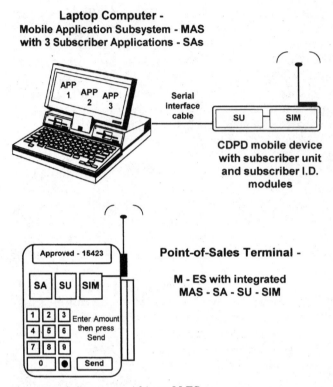

Laptop Computer -
Mobile Application Subsystem - MAS
with 3 Subscriber Applications - SAs

APP 1 APP 2 APP 3

Serial
interface
cable

SU   SIM

CDPD mobile device
with subscriber unit
and subscriber I.D.
modules

Approved - 15423

SA SU SIM

1 2 3   Enter Amount
4 5 6   then press
7 8 9   Send
0 ● Send

Point-of-Sales Terminal -

M - ES with integrated
MAS - SA - SU - SIM

Figure 9-1 Subsystems within an M-ES.

personal identification numbers, etc. Figure 9-1 shows these relationships.

Let us first consider interfacing IP hosts in non-CDPD networks. Figure 9-2 shows an IP host connected via an Ethernet or other local area network. Installations similar to this require the use of a device driver to mate the IP host with the *network interface card,* or NIC (not to be confused with "network information center"). The device driver that is used defines the network type, media access, and level 2 framing specifics. Recall that CDPD mobile devices should be able to interface with existing applications with little or no modification. CDPD SUs, therefore, should have the appearance of a network interface card to the MAS entity, as illustrated in Fig. 9-3. With this in mind, a CDPD device which is either integrated into a PC platform or of the PCMCIA variety should be compatible with the major types of device drivers used in the industry. Driver types include NDIS (Microsoft), ODI (Novell), and Apple varieties. Other driver types, such as Clarskson drivers, may also be supported.

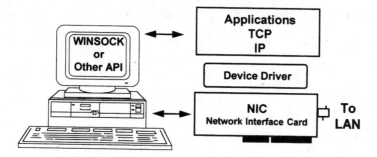

Device drivers enable application "stacks" to
communicate with network interface hardware.

Figure 9-2 An example of elements required to connect a host to a local
area network.

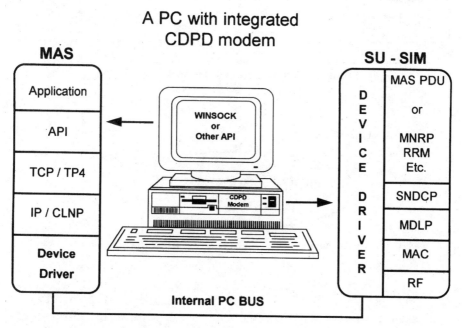

Figure 9-3 Integrated CDPD modems should have the appearance of a network interface card to its host computer.

SLIP / PPP access with external wireline modem

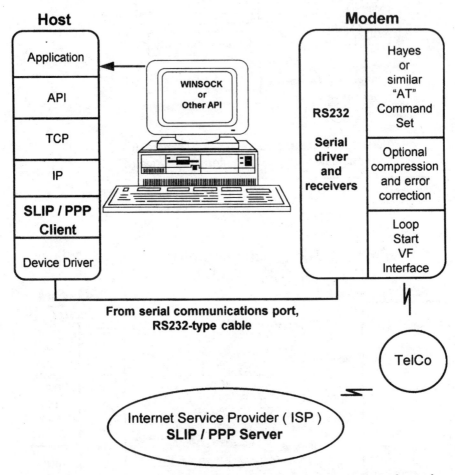

Figure 9-4 Block diagram of a typical Internet access arrangement using modem technology.

Figure 9-4 shows an IP host communicating via serial cable. A typical scenario for an IP host communicating over serial interfaces includes the use of either SLIP (serial line interface protocol) or PPP (point-to-point protocol). Note in this diagram that the IP host is communicating with a network access "server." SLIP/PPP network access servers typically reside at a point of presence (POP) to the Internet (or other) network environment. The POP is either at the remote endpoint of a private line using DSU technology, or a dial-up connection using ISDN, SW 56 (switched 56-kilobit service), or, as this figure shows, modem technology. The physical placement of the user clients and network servers changes somewhat when using SLIP or PPP to access the CDPD network.

Just as with a dial connection, CDPD access via SLIP/PPP is established between the application subsystem and the CDPD POP. However, the CDPD "point of presence" is locally attached to the user application subsystem, as opposed to being found at a remote serial line endpoint. In this mobile environment, the M-ES subscriber unit itself is the CDPD network POP. This makes sense if one takes into account that DSMA/CD, MDLP, MNRP, etc., are all specific functions of the CDPD network, which are resident in the SU. Therefore, the "network" extends out to the user and terminates at the SU. So, while the "logical" view of a user accessing a CDPD network via SLIP/PPP is equivalent to terrestrial implementations, the "physical" view changes. As Fig. 9-5 shows, this is because the CDPD network "comes to the user," whereas in terrestrial networks the user "comes to the network." This is the manner in which most mobile IP hosts will interface with the CDPD network. User PCs will utilize serial connections with an M-ES which supports SLIP and/or PPP. This configuration completely hides the details of a wireless connection to the internet. Carrier-provided IP addresses must be used, but other than this requirement, no other changes need to be made at user applications.

As is the case with terrestrial SLIP/PPP communications, each endpoint of the serial connection has an IP address associated with it. The serial line protocols negotiate various working attributes such as security and compression with each other before actual communications begin. Depending on the serial line protocol used, each endpoint may be able to dynamically learn the remote endpoint's IP address, or the IP addresses may be configured in a static manner.

Because the addresses associated with the serial link have significance only between the locally attached MAS and SU entities, "server" addresses can be duplicated. A CDPD carrier assigns a single IP address, which is duplicated for each SU resident SLIP/PPP server. The server address is a single address that is unique to all M-ESs within a given home area. Because SLIP server addresses are not known to the network, the address itself does not have to be a "real" one. Cincinnati Microwave, a manufacturer of a very popular M-ES, typically uses the address 1.1.1.2 for the SLIP server. This relationship is shown in Fig. 9-6. The user's application system as well as the SU entity's NEI are unique for each M-ES and therefore are not duplicated. Both the SU and server addresses are assigned to the user by the CDPD carrier administration. Some addresses assigned to a user by the carrier may appear twice: within the user's MAS as an IP address, and within the user's attached mobile device as an NEI.

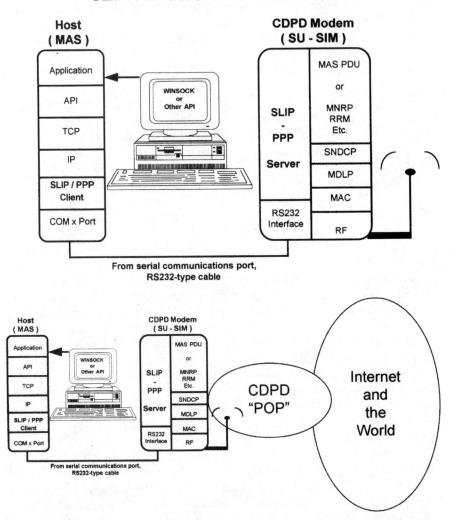

Figure 9-5 With CDPD, a SLIP/PPP server is a local function residing within the end user's mobile equipment.

The SU address is used for registration purposes. Some mobile devices can be programmed with multiple NEIs, each NEI, for example, belonging to different network service providers. If multiple NEIs are being used by an M-ES, it is possible that the MAS IP address and SU NEI are different. This is the case with terminal D in Fig. 9-6. Note that while the users' home MD-IS must perform the authentication and registration tasks required, the user does not know the

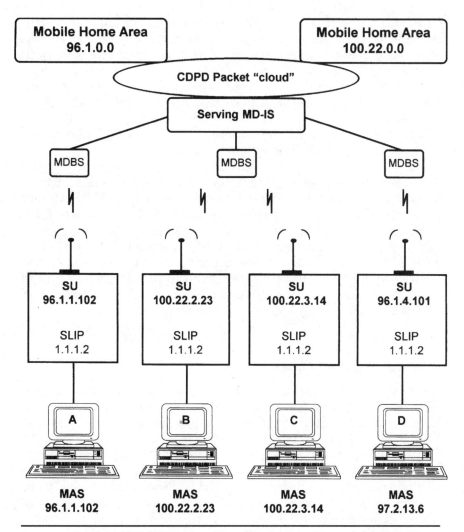

Figure 9-6 A unique SLIP/PPP server address is not required for each mobile end user. However, the address used for registration purposes (SU/NEI) is unique and must pass authentication procedures.

address of the MHF. The actual address of the home function is maintained within serving MD-IS directories. M-ES and home MD-IS (sub)networks are used to determine "what" MHF each mobile device should be associated with. This is also illustrated in Fig. 9-6. Notice the four NEIs are members of two different serving areas, or subnets.

Maintaining coherence with existing IP standards and accepted connectivity methods is important for the CDPD mobile device manufacturer. As long as a user's application subsystem can connect to a

CDPD SU via SLIP, PPP, or industry-standard device drivers, application developers will have no trouble making use of the mobile network. Most networking applications as they now exist are already IP compatible. There is a tremendous wealth of programmers who already know how to "port" applications to an IP environment. Accordingly, finding the talent to write CDPD applications should not be a high hurdle to overcome. As mentioned earlier, the challenge is to introduce streamlined applications that will not be hampered by the bandwidth limitations of cellular channels.

Many M-ESs have a TCP and UDP interface resident within the SU itself. An applications programmer can take advantage of well-understood "porting" procedures to connect to such an M-ES without requiring a PC or laptop-type machine as a host. An advantage of this feature is that small, low-cost machines with limited processing capability can take advantage of the CDPD network. Meter-reading devices, for example, may fit into this category of end system. Mobile devices with TCP/UDP interfaces give the programmer a way to choose the type of transport required (connectionless or connection oriented), as dictated by application requirements.

Character-Based Interfaces

Many applications expect to communicate via asynchronous AT-compatible modems as opposed to an IP stack. In order to accommodate these needs, a CDPD-specific variation of the popular AT command set has been defined. Many of the commands that are understood with the Hayes AT command set are used with the CDPD command set. Parameters used with these commands, however, may be unique to CDPD. An example is the dial command ATD. Normally, telephone digits accompany this command. CDPD ATD commands, in contrast, use IP addresses in standard (example: 191.254.3.10) "dotted decimal notation."

When employing the CDPD AT command set, a user application is actually interacting with a TELNET client resident within the mobile's SU. Use of TELNET implies use of TCP/IP. Asynchronous keyboard information is passed from any terminal emulation function (such as the Windows "terminal") and handed to TELNET via the AT interface, as illustrated in Fig. 9-7. The TELNET service functions as a packet assembler/disassembler (PAD).

As with any packet assembler and disassembler, the PAD function needs to know how to treat data presented to it. Some obvious data format essentials, such as speed, intelligence bits, parity, and stop bits, must be defined. Flow control mechanisms employed between the MAS and SU must be defined. Data forwarding conditions,

CDPD Access with Terminal Emulation

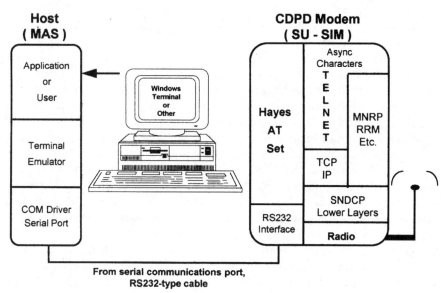

Figure 9-7 CDPD M-ES equipment incorporating TELNET clients to accommodate attached hosts using terminal emulation.

local/remote echo definitions, and many other parameters must be defined. These parameters are defined via the CDPD command set. Perhaps not so obviously, CDPD-specific parameters need to complement those that would normally be defined in a PAD environment. The *register* and *deregister* commands are examples of CDPD-specific commands. User security features may also be programmed into the SU, such as a PIN (personal identification number), which will enable the device for use. Three main modes of operation must be supported in order to provide these services:

- *Administrative mode.* Two subsets apply to the administrative mode. The first is a mode, where hardware and operational behavior can be defined. Examples include flow control parameters and data forwarding definitions. The second subset includes user-specific information such as NEIs which have been assigned and personal identification numbers.

- *Command mode.* The mode of operation where data that is presented to the SU is not forwarded as user data packets, but interpreted locally as a command to dial, hand up, register, etc.

- *Data transfer.* Information presented to the SU is packetized as shipped as per data forwarding definitions made while in the command mode.

When making use of the AT interface of an M-ES, it is important that data forwarding definitions be defined in a sensible manner. It is undesirable to transmit individual characters over an air link and through the network utilizing remote echo. For one thing, forwarding individual characters as they are typed via TELNET is terribly inefficient. Each character has to be encapsulated within TELNET, TCP, IP, SNDCP, and MDLP. The majority of precious bandwidth will be consumed by protocol overhead information, not actual user data. If the service is a metered service, customers will be paying for transport of this overhead, thus making remote echo expensive in real dollar costs as well. User satisfaction will also suffer. Using remote echo in a store and forward packet mode system will cause delays in character presentations on users' screens. For these reasons, an SU should be configured for local echo, where the user is able to view characters as they are typed. These characters will be buffered by the SU until a data forwarding condition has occurred. The remote system should be configured in such a way as to be able to handle "block mode" data input. This approach will increase user satisfaction, conserve bandwidth, and minimize the costs associated with use of the network. Remember that from the carriers' point of view, all data transported is billable, and this includes protocol control information that accompanies actual user data.

Criteria for forwarding data packets in the CDPD AT environment is similar to any other PAD environment, such as in an X.25 implementation. User data is collected as it is presented to the M-ES's subscriber unit. When the (user-defined) forwarding condition occurs, a packet is placed within an MDLP frame and transmitted out over the reverse channel. Forwarding conditions can include filling up a complete packet (130 bytes minus PCI information), a special user-defined forwarding character, or an intercharacter time-out period. The intercharacter time-out period defines how long the PAD function will wait for additional characters before transmitting data currently being held in its input buffer. This function may come into play between keystrokes if a human operator is interacting with the SU. In such a case, it is important to define the intercharacter time-out period to be long enough to allow for pauses which will naturally occur when a human is either a slow typist or is distracted. Disabling this function and forcing the user to input a special data forwarding character ensures that only data intended to be transmitted is transmitted. When defining the data forwarding character, the user must determine if this character should be forwarded with the data itself, or omitted from the transmitted packet.

CDPD AT Commands and Register Sets

The following is a summary of "basic" CDPD-defined AT command functions.

> AT or A/   Attention command while in command mode. Carriage return or user-defined contents of S3 register exits command mode. Space characters may be used to separate concatenated commands on the same line. "Attention" (AT -A/) must be uppercase. Autobaud, if enabled, will operate on attention inputs.
>
> ATA   Enables SU to answer a TELNET connection request. This condition lasts for S7 (register) seconds.
>
> ATD(string)   Dial command. The (string) is an IP address followed by an optional TCP port number (23 is the TELNET default). Example: ATD192.254.3.1/23.
>
> ATE(0/1)   0 = No command mode echo. 1 = Command mode echo enabled.
>
> ATF(0/1)   0 = Local echo. 1 = No local echo.
>
> ATH(0-3)   0 = Transmit assembled data and hang up. 1 = Return service code. 2 = Purge assembled data and hang up. 3 = Purge assembled data, hang up, and deregister from network services.
>
> ATI(0-3)   Identify: 0 = EID, 1 = firmware version, 2 = manufacturer, 3 = model number.
>
> ATO   Returns to on-line/data transfer mode.
>
> ATQ(0/1)   0 = Do not send service codes to MAS. 1 = Send service codes to MAS.
>
> ATS(register number)?   Reads contents of register.
>
> ATS(register number) =   Sets contents of register.
>
> ATV(0/1)   Defines whether service codes are numeric or alphabetic. 0 = Numeric. 1 = Alphabetic. Service codes are defined as: 0/OK, 1/CONNECT, 2/RING, 3/NO CARRIER, 4/ERROR, 6/NO DIALTONE, 7/BUSY, 8/NO ANSWER.
>
> ATX(0/2)   0 = Extended service codes apply for 0 through 4. 2 = Enable all extended service codes.
>
> ATZ   Releases any active connection but maintains registration status with network. Resets all parameters to saved profile.
>
> AT&C(0-3)   0 = DCD always active. 1 = DCD follows connection establishment state. 2 = Always on but flashes at disconnect. 3 = DCD on when "RF in range."
>
> AT&D(0-2)   0 = Ignore DTR. 1 = Enter command mode when DTR toggle from on to off. 2 = Release connection when toggle to off is detected. Enable answer mode only when DTR is on.
>
> AT&F   Resets SU to factory defaults.

AT&S(0-2)   0 = DSR always active. 1 = DSR toggles with connection established state. 2 = DSR toggles with condition of battery.

AT&V   View profile in use.

AT&W   Writes current configuration into NVRAM to become the new active profile.

AT&Z(n)(string)   Attaches (string) destination address to (n) value for future dialing purposes.

An extended set of the CDPD AT command set is used to control PAD functions and attributes associated directly with data format and treatment. Following is an overview of these extended commands.

AT&Lb,d,p,s   Sets baud rate, data (intelligence) bits, parity, and stop bits of MAS–SU interface.

AT\A   A mobile device may be able to operate in the circuit switched AMPS environment as well as the CDPD environment. AT\A switches the operating mode to AMPS.

AT\F(0-3)   0 = Do not include data forwarding character in assembled packet. 1 = Register S51 forwarding character is included in assembled packet, but register S52 forwarding character is not. 2 = S52 is included, but S51 data forwarding character is not. 3 = Both data forwarding characters are included in assembled packet.

AT\M(0/1)   Overrides the previous AT\F command. 0 = Always include forwarding characters in assembled packet. 1 = Include forwarding characters in assembled packet as per AT\F command settings.

AT\N(3/4)   3 = Requests A side carrier. 4 = Requests B side carrier.

AT\P(+,-, =)   Enable PIN. (+) Disable PIN. (-) Define new PIN. (=) Syntax for new PIN definition is as follows: AT\P = [current value],[new value],[new value].

AT\Q(0-3)   0 = No flow control over MAS–SU interface. 1 = Xoff/Xon. 2 = RTS/CTS hardware flow control is used. 3 = Both forms of flow control are used.

AT\R(0-2)   0 = Deregister. 1 = Register. 2 = Automatically register with connection request (ATD) and deregister with release (ATH).

AT\S(n,?)   Defines address (n) to use for registration. To view NEI list, use (?).

AT\T(0,1)   0 = No intercharacter time-out period before packet forwarding. 1 = Refer to register S50 for time-out period in 1/10-second increments.

An M-ES subscriber unit's configuration is stored in a set of SU (S) registers. The entire set of registers, when viewed collectively, is referred to as a *profile*. Similar to other PAD environments (such as X.25), an SU may have more than one profile set defined. One profile set will always define factory default values. Recalling this profile set provides a good way to undo configurations that have been repeatedly modified, allowing new definitions to commence from a "clean slate." At least two additional profiles will exist. The active profile is the set of definitions that are in use and defining the operational characteristics of the M-ES at the moment. The active profile is used as long as power is not interrupted, or the user enters command mode and makes changes to the active profile set. A "saved" profile also exists. This is the profile set that is used when a power interruption occurs. It is not uncommon for a PAD device to enable storage of multiple saved profiles. Having the ability to choose saved (named) profiles for use enables the user to tailor the operating environment for specific application requirements. Figures 9-8 and 9-9 provide a listing of CDPD S registers.

CDPD M-ES Subscriber Unit S Registers

Register Name	Possible Values	Comments
S0	0,1	Default auto answer off = 0
S2	IA5 characters	Escape code default = " + " decimal 43. Enters SU into command mode.
S3	IA5 characters	Defines carriage return character. Default decimal 13.
S4	IA5 characters	Defines line feed character. Default decimal 10.
S5	IA5 characters	Defines backspace character. Default decimal 8.
S7	Second increments	Wait for connection establishment. Default 60 seconds.1
S12	In 1/50th seconds	Escape code timer. Default 50 = 1 second.
S14	Read-only bitmap	Bit 0 = F command setting. Bit 1 = E command setting. Bit 2 = Q command setting Bit 7 = &E command setting.
S21	Read-only bitmap	Bits 1,2 = DSR setting - &S Bits 3,4 = DTR setting - &D Bits 5,6 = DCD setting - &C Bit 7 = Escape mode setting - &E
S22	Read-only bitmap	Bits 4,5,6 = Extended service codes / X command setting.
S23	Read-only bitmap	Bits 1,2,3 = Baud rate. 0/1200, 1/2400, 2/4800, 3/9600 4/19200, 5/38400, 6/57600, 7/115200 bps. Bits 6,7 = Parity 0/even, 1/none, 2/odd, 3/mark

Figure 9-8 Subscriber S registers.

CDPD M-ES Subscriber Unit S Registers - Continued

Register Name	Possible Values	Comments
S50	In 1/10 second increments.	Data forwarding timeout. Default = 20 - seconds.
S51	IA5 characters	Primary data forwarding character. Default = [CR].
S52	IA5 characters	Secondary data forwarding character. Default = [cntl-Z].
S53	String	Remote address (read only).
S54	0 - 255	Reverse channel packet queue (read only).
S55	0 - 255	Forward channel packet queue (read only).
S56	0 - 255	Extended network error code (read only).
S57	Read-only bitmap	Bits 0,1 00 = M-ES unregistered. 01 = M-ES registered.
S58	Read-only bitmap	Bit 0 Auto transmit \T command setting. Bit 1 Manual transmit \ M command setting. Bits 2,3 MAS flow control \Q command setting. Bits 4,5 Data forwarding \F command setting.
S59	0 - 255	Number of seconds to sleep. Default = 0.

Figure 9-9 More S registers.

Tuning Applications for Mobile Environments

We can now see that connecting applications to the CDPD network in and of itself is not a difficult task. For applications that run on top of existing TCP/IP stacks, SLIP and PPP provide a simple way to interface with a CDPD modem. These applications, such as Netscape or other World Wide Web browsers, require no modification in order to interface with CDPD. For applications that will reside on machines with limited processing power, or machines that can only support terminal emulation, mobile devices offer TELNET and TCP/UDP interfaces to which a programmer can "port."

While interfacing applications as they exist today will work with CDPD in the sense that the user gains the ability to communicate, doing so may be an expensive proposition. In this context, "expensive" has multiple meanings. One interpretation of expensive is the inefficient use of the cellular air link, which may contribute to (medium) access delays, throughput degradation, and increased response times. Another interpretation of expensive has to do with real dollars being spent as a result of the amount of data being passed over the network.

Remember that many applications, especially TCP/IP applications, were never intended to be used in a network environment where usage-based billing was an issue. Also, these legacy applications were written with the assumption that high-speed local area networks would be the transport medium, and internetworking would be accomplished via leased lines. Creating an application to communicate in an efficient, bandwidth-conserving manner was not a primary concern for many designers when millions of bits per second of (local) transport was available and network costs were assumed to be fixed.

Consider Novell's very successful network system, Netware. The original version of this product's internet packet exchange (IPX) network protocol is a good example of a networking system which needed to change with the times. This protocol was a half-duplex protocol with no windowing capability. Transferring even large files of 300 kilobytes worked so well that responses seemed instantaneous, as long as the communications were confined to a local area Ethernet segment. However, when communicating via the "wide area," even at 64 kilobits per second, response times and throughputs were affected dramatically. With the success of the product, user needs had changed. Geographically dispersed workgroups that operated among themselves soon had a need to access information in remote offices. With this new requirement to communicate over long distances, there was a need to modify IPX's behavior. As a result, Novell introduced a version of its network protocol called "burst mode," which was better suited for wide area operation and included windowing capabilities. This is just one example of how a popular product adapted to new demands resulting from changing network patterns. We have all seen that as networking technologies change, applications must be modified to cope with increasing and unanticipated user demands. With the introduction of CDPD, existing applications will need to undergo similar modifications to meet user expectations, and new applications should recognize the attributes of the wireless network supporting them (limited bandwidth, usage sensitive).

Let's take a look at some ways to "streamline" a process that runs on top of the CDPD network. Employing data compression mechanisms within the MAS entity is one way to "buy back" transmission efficiency, and perhaps even reduce the billable amount of characters sent. Remember that CDPD's encryption and compression features operate only over the air link. SNDCP within the MD-IS will re-create original text streams before the metering process is encountered, as illustrated in Fig. 9-10. What this means is that CDPD's data compression feature is intended to reduce the amount of data transmitted over the air interface, not reduce the billable traffic (revenue) presented to the network. Exposing user data to a compression process before the M-ES/SU interface will reduce bandwidth consumption and usage fees. The drawback to this, however, is that the remote end

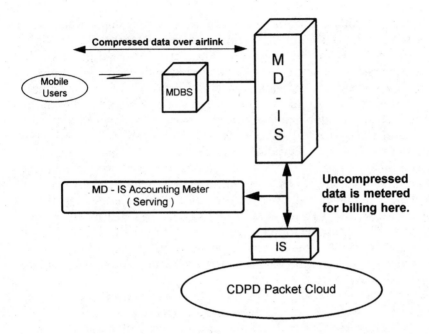

CDPD data compression enhances airlink efficiency but does not reduce billable characters sent through the network.

Figure 9-10 CDPD compression will not reduce the amount of data perceived by the carrier to be billable traffic.

of the communications process will need to decompress the data. For custom applications, this may not be a problem; in fact, it is desirable. However, when communicating with other (open) systems, compression mechanisms must be bypassed. An example of this scenario is a mobile user checking inventory in a remote database. The user's database application incorporates compression. As this is the primary application used, savings are realized. On occasion, the user connects to web sites and other Internet resources. During these sessions, compression cannot be enabled, and usage costs rise.

This is not to say that CDPD should not be used for Internet access. There is a difference between Internet "access" and casual "browsing." It is fairly well accepted that "casual browsing" in a metered network environment, CDPD or wireline, can be expensive. One is better off accessing the Internet for entertainment purposes via an ISP (Internet service provider) that offers attractive monthly rates, using a high-speed modem. Internet "access," on the other hand, need not be prohibitive in terms of expense. Access to the Internet has real and significant business value. Having real-time access to electronic mail services, corporate web servers for pricing and product information, and public databases for industry-specific information is increasingly becoming a tool that business users cannot do without. There are many small things that a user can do to minimize time spent and characters transmitted while accessing the Internet when there is a requirement to go "on-line." Users need not "surf." Users should think of the Internet as a tool. For real business use, create a file or hotlist with all the pointers and addresses that are frequently accessed. Turn off graphic capabilities. While graphics and sounds are pleasing, most of the usable information which has real value is in text form. With planning, the user can minimize traffic generated that has little or no value, and reduce traffic generated during searches by using scripts and hotlists. To illustrate this point, a case study is offered.

A technical professional uses the CDPD network primarily to access corporate databases which contain information related to products and services that the company provides. While at a customer's facility, a requirement arises for information that is accessible through the Internet. The information of interest consists of technical information on the wide area network technology, frame relay. Accordingly, the user connects to the Frame Relay Forum's World Wide Web server to retrieve the desired documents.

Appendix F in this book contains two traces of IP packet flow created when connecting to this web server. In the first trace, we see the user access the Frame Relay Forum's "home page" at the URL:

http://frame-relay.indiana.edu/

From the home page, two submenus are accessed via pointers:

http://frame-relay.indiana.edu/frame-relay/5000/5000index.html

and

http://frame-relay.indiana.edu/frame-relay/5000/5001.html

Finally, the user finds an area which contains "approved documents" relating to the technology of interest. The trace of this session shows the IP packets that are created as a result of the user browsing for this information.

In the second trace, we see the results of a more organized search. The assumption here is that the user had at one time previously accessed the same page defined by the final URL:

http://frame-relay.indiana.edu/frame-relay/5000/5001.html

Knowing that approved documents on this technology can be found at this pointer, the user has created a hotlist which contains the pointer and a clear description of what can be found there. When information on this subject is required, the user can go directly to the depository of the data and bypass irrelevant pages. Because the user bypasses unwanted information, much less data is generated and passed over the air interface, resulting in more digestible usage fees. Of course, occasional browsing of on-line databases will always occur. With planning, however, occasional internet access via CDPD need not be a prohibitively expensive proposition.

With respect to network protocols, TCP offers a reliable transport mechanism for mobile applications. However, TCP is a "chatty" protocol which causes many datagrams to traverse the air link. A programmer writing custom applications can embed integrity checks and error recovery routines in the application that are more efficient and cost effective than TCP. These applications can then be ported to UDP to minimize traffic generated over the air link. TCP acknowledgments are also billable, and they occur frequently. It is desirable to design in such a way as to minimize reverse-path data flows due to acknowledgments. If successful, customer billing will reflect more the cost of transmitting information as opposed to the cost of managing information transfer. A scenario is offered: A neighborhood patrol security application requires the security officer to indicate that certain checkpoints have been reached during the tour of duty. When the application acknowledges the visit to a checkpoint, the security officer may continue the tour. On the surface, it appears that two transmissions are generated, one checkpoint indicator created by the officer, and an acknowledgment for the officer's transaction. However, because TCP is used as a transport mechanism, TCP generates a layer 4 acknowledgment for each application PDU transmitted. In this scenario, a TCP acknowledgment is sent for the officer's original checkpoint indicator. The result is that two (billable) packets traverse the air interface. As the application itself acknowledges the officer's indication, another TCP level acknowledgment is generated. Again, two billable packets traverse the air interface. Thus, using TCP in effect doubles the CDPD traffic sent, which doubles the customer's bill. Using UDP and concatenating application-level "acks" with response PDUs will minimize the traffic sent, and reduce the customer's bill. Housekeeping activities such as sequencing at the application level can compensate for the unreliable nature of a UDP transport service. Figure 9-11 illustrates this point. Another possible strategy to reduce traffic flow over the air interface is to utilize a "shifting" acknowledgment scheme. As datagrams are transmitted, an application or transport service may acknowledge datagrams at an increasingly wider interval. An "end of segment" or "end of message" indicator can be used to solicit an acknowledgment when there are no more datagrams

Using TCP over Air Interface

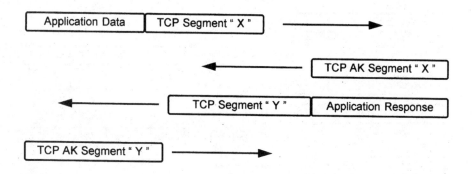

Using UDP over Air Interface

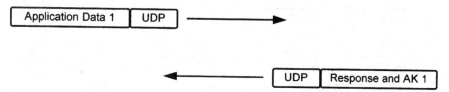

Figure 9-11 Incorporating UDP and application-level acknowledgments will help to reduce air link loads and usage fees.

to transmit. Resequencing of missing PDUs can be handled by a reject mechanism similar to what is supported by the MDLP. This is illustrated in Fig. 9-12.

Some Thoughts about Security

Security is always a concern. This is true regardless of whether the computing is being performed locally, or remotely via a communications network service. Because of their wireless nature, it is understandable that many users have a greater reluctance to implement RF broadcast technologies such as CDPD when choosing a network service. While some fears are real, and worth consideration, others are quite frequently overblown. When considering the issue of net-

Shifting Acknowledgments

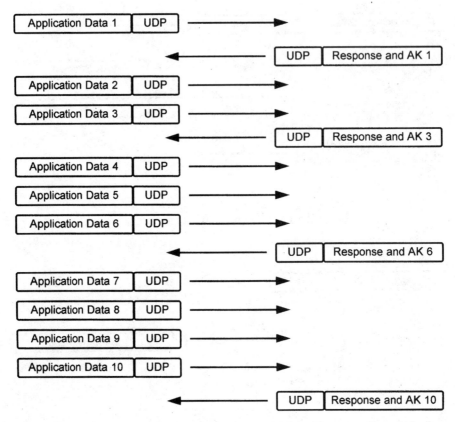

Figure 9-12 An example of an attempt to reduce billable traffic is seen here with the incorporation of an acknowledgment scheme which reduces the frequency of "AK" generation.

working and security, the user needs to be able to identify what type of security breach needs to be addressed with a defense mechanism. Simple eavesdropping can result in theft of information, including passwords as well as user application data. Unauthorized access to a system can result in not only information theft, but damage to the system in the form of data corruption and system unavailability.

Security concerns should be applied to each of five separate levels:

- Applications themselves
- User hosts and servers
- Network-level switching and routing infrastructures

- Physical media
- Bit-level transmission

Let us first consider bit-level transmission. The only way to protect the confidentiality of data, regardless of the type of data being transmitted, is to encrypt the data transmitted. Protecting the privacy of network-level addresses requires encryption of source and destination addresses. This is perhaps easier said than done, as doing so requires the implementation of network-wide switching and routing technologies that do not exist. Accordingly, data encryption is typically limited to the user payload portions of packets being sent through a network environment. Network-level header information such as source and destination addresses are not modified, as encrypting these fields would prevent a switch or router from forwarding packets properly.

The CDPD network will work toward keeping user data confidential, but only over the air interface. The combination of header compression, dynamically changing encryption keys, and actual encryption of user payloads will be sufficient to protect the users' data as it traverses the air interface. Users may wish to take additional steps to ensure the privacy of their data on network segments other than the air link. If this is the case, the most sensible approach is to ensure privacy at the application level. This implies the use of a client–server type of relationship. The client and the server need to cooperate with each other in both encryption and key management services.

With respect to securing physical media itself, the CDPD network is perhaps vulnerable to "Captain Midnight"–type intrusions. "Captain Midnight" gained notoriety some years ago by the criminal act of pirating a satellite channel and broadcasting program material that was of interest to himself, but few others. The captain was later fined and served jail time for his deeds. Vulnerability of the air interface exists in that a criminal may broadcast over cellular frequencies, causing a jamming effect. However, any intrusion of this type would be limited to a single channel or set of channels within a particular cell; the remainder of the network would function normally. Also, the channel-hopping capabilities of an M-ES might be able to compensate for interference by simply moving to an acceptable channel. Network terrorism such as this is a possibility, but remote. One may argue that acts of network terrorism would more likely be directed toward the company or organization making use of the CDPD network, perhaps by a disgruntled worker. The real concern in this scenario is making sure that hosts, wiring rooms, and software are secured at the customers' locations.

Attempts at unauthorized access to the CDPD network is a more realistic scenario. This is prevented by the security management protocol and authentication procedures that are in place. Recall that the M-ES, serving MD-IS, and home MD-IS all cooperate by exchanging and relaying authentication credentials. With each registration, new credentials are agreed upon and subsequently transmitted to mobile server functions in encrypted form in new ESH messages. A "hacker" wishing to gain unauthorized access to the network via CDPD channel streams would have a very difficult time of it, to say the least. The odds of successfully breaking these codes are very low. The stolen information would then have to be put to use before the next valid registration occurs, because at that time a new set of valid credentials would be generated, making the stolen credentials useless.

Making network-level switching and routing devices secure is of paramount importance for both the network services operator as well as the user. This is an issue deserving of consideration from both the user and carrier communities. Access lists and "firewalls" should be constructed so as to prevent the pass-through of packets from unknown sources. Network-level switching elements also need to be protected against invasion via remote access programs such as TEL-NET. How are unauthorized packets prevented from using the link between users' fixed-end systems and a CDPD POP? Can any party on the network "ping" (a standard continuity test) a mobile-end system? The industry-standard "ping" command allows users to define how much data will be transmitted and echoed back, and also define how many repetitions (including "forever") should occur. Who pays for the use of the air link if a third party pings a mobile device with malicious intent? Users and network services providers should address these matters during the initial discussions of a service agreement.

Users should consider using a challenge-and-response access method to supplement password authentication on their network servers and hosts. A simple example of this being applied is shown in the scenario illustrated in Fig. 9-13. A mobile application may utilize a set of credentials consisting of serial numbers, MAC addresses, employee numbers, or similar private information. This information may reside within different credential sets at the mobile-end system. Customer servers may utilize an index which references mobile user names or passwords and the information contained within each credential set. As shown in Fig. 9-13, there are four credential sets for this M-ES. The challenge-and-response system is very simple, but substantially more secure than simple password authentication alone. When the mobile user is prompted for a password, the server may ask for the information contained within a specific credential set. During subsequent log-on attempts, the server may request the information contained within the next sequentially numbered credential set. In this manner, the credential sets used for each log-on attempt will vary in a cyclic fashion.

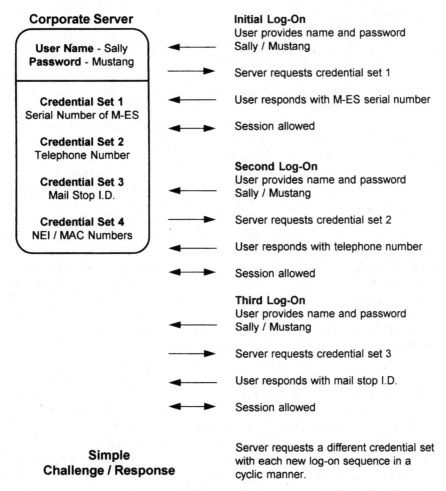

Corporate Server

User Name - Sally
Password - Mustang

Credential Set 1
Serial Number of M-ES

Credential Set 2
Telephone Number

Credential Set 3
Mail Stop I.D.

Credential Set 4
NEI / MAC Numbers

Initial Log-On
User provides name and password
Sally / Mustang

Server requests credential set 1

User responds with M-ES serial number

Session allowed

Second Log-On
User provides name and password
Sally / Mustang

Server requests credential set 2

User responds with telephone number

Session allowed

Third Log-On
User provides name and password
Sally / Mustang

Server requests credential set 3

User responds with mail stop I.D.

Session allowed

**Simple
Challenge / Response**

Server requests a different credential set
with each new log-on sequence in a
cyclic manner.

Figure 9-13 A crude example of a challenge-and-response access protection mechanism.

Ultimately, it is an application that is being accessed and used. Multiple applications may reside on any given network server. For this reason, security mechanisms should be written into the applications themselves. In this way, each application may have its own unique security measures invoked. Once successful log-on to a server is accomplished, individual applications can be protected using encryption, challenge-and-response handshakes, or similar mechanisms.

Users should always keep in mind that the CDPD network is an open, standards-based network. While substantial efforts have been made to create a reliable and secure service, especially with respect to authentication and the air interface, users should protect their own interests and create secure applications when the need exists.

The CDPD Hybrid Service

In an effort to complement the coverage area of CDPD service, a "hybrid" service has been defined by the CDPD Forum called *circuit switched CDPD* (CS-CDPD). With the availability of CS-CDPD, a mobile user will be able to access the CDPD network via telephone line and circuit switched cellular calls. The idea behind the service is simple; provide user connectivity to a common set of internet resources via any available access mechanism, packet mode or circuit switched.

New terminologies and devices are associated with the service. A mobile device which is capable of using the CS-CDPD service is called a CM-ES, or *circuit-mode mobile-end system*. An MD-IS which supports the service is referred to as a CMD-IS, or *circuit-mode mobile data intermediate system*. A CS-MB, or *circuit switched mobile base station,* serves as a modem bank for calls being placed by CM-ESs.

Figure 9-14 is a block diagram of the network components. Notice that a new process is involved, the CSCCP, or *circuit switched CDPD control protocol*. This layer sits just underneath the SNCDP process which is part of packet-mode CDPD operations. Looking upward, the CSCCP appears to provide the same services that MDLP provides to the convergence protocol. Looking downward, CSCCP is aware of circuit switched environments and is not concerned with the point-to-multipoint link control issues that the packet-mode version (MDLP) is concerned with.

Obviously, the first step to using the service is to place a call. Once a circuit switched call has been placed, the two modems involved must synchronize with each other and maintain a physical connection. All modems associated with the CS-CDPD service must support (at a minimum) the V.32, V.32bis, V.22, and V.22bis modulation schemes. After achieving synchronization with each other, the modems must establish a data link control dialog. The data link control dialog used over the circuit switched connection may be one of the commonly found link control mechanisms in use today, such as V.42, ETC, or MNP10.

After establishment of a data link, the CSCCP entities at either end of the call must initiate a series of message exchanges which will enable the mobile device to use the CDPD network. At this moment, the connection between a CM-ES and the serving CMD-IS is point to point. Because the connection is point to point in nature, a simple framing mechanism can be utilized. It has been agreed upon that the framing mechanism for data PDUs relayed between CM-ES and CMD-IS CSCCPs will use the industry-standard SLIP as defined in RFC1055. SLIP is an extremely simple protocol. It is not concerned with addressing, error detection or correction, or any other control mechanism. It is merely a framing protocol which identifies the beginning and end of a message which is carrying IP (or other) packets.

The SLIP defines two special characters: END and ESC. The END

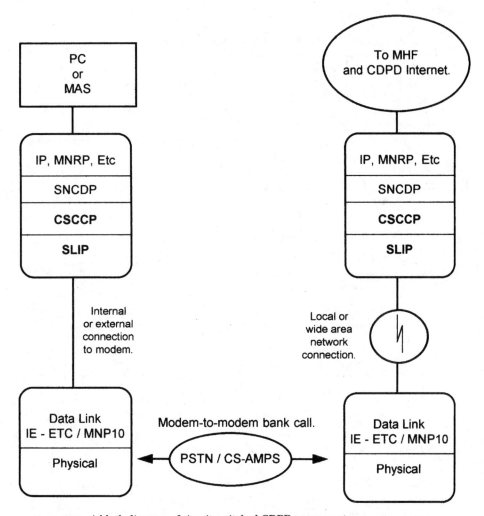

Figure 9-14   A block diagram of circuit switched CDPD components.

character is a hex C0 (decimal 192). A SLIP ESC is a hex DB (decimal 219), which is different from the ASCII escape character. When transmitting frames, SLIP simply starts the frame with an END character, sends the contents of the message, and finishes the frame with another END character. If part of the transmitted data looks like the SLIP END character, a 2-byte code, [SLIP ESCAPE-DC hex], is sent. If part of the data within the frame looks like a SLIP ESCAPE, the 2-byte pattern [SLIP ESCAPE-DD hex] is sent. When the last byte in the packet has been sent, an END character is transmitted, signifying the end of frame. Figure 9-15 shows the encapsulation used between CSCCPs.

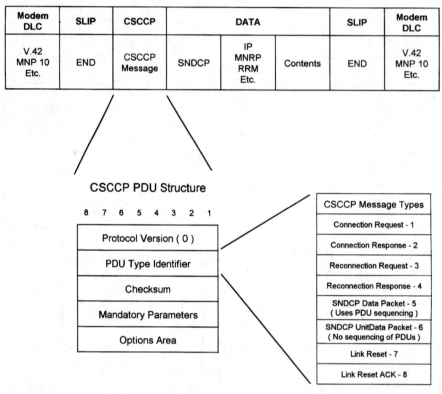

Figure 9-15 Encapsulation methodology of circuit switched CDPD.

The CSCCP entities must now connect to each other. The mobile unit issues a connection request message and awaits a connection response message. If there are adequate resources to handle the connection request, a result code in the connection response indicates "successful." Because CSCCP emulates the mobile data link protocol, CSCCP indicates that the data link layer is "up" via the "data link established" primitive (DL_Establish_IND). Security management processes then begin, and when the SME encryption processes are completed, the mobile registers with MNRP by normal procedures.

One of the very interesting features of the CS-CDPD service is its "redirecting" capability. Because CDPD is "overlaid" on top of the existing AMPS infrastructure, a given CDPD area can be associated with a given AMPS system identifier. If this is the case, it is possible to provision the service in such a way that a CMD-IS is aware of local access numbers based on AMPS system identifiers. Accordingly, when dialing via the AMPS, a mobile device will include an AMPS system identifier within the connection request message. CMD-ISs handling incoming calls can refer to a database which indicates local access numbers to use based on AMPS system IDs. If it turns out that a mobile user is placing a call to a distant CMD-IS, the called system can redirect the mobile to another access line's telephone number. By informing a mobile to call another number, users can minimize unwanted expenses due to long-distance charges. AMPS system IDs and redirection numbers to call are included in various option fields within CSCCP packets. An example of redirection in action is provided in Fig. 9-16.

Notice in this example that the Chicago metropolitan area has two CMD-ISs in service, but the user calls long-distance to a Detroit CMD-IS. The called system in Detroit can do more than simply inform the user to redirect toward a local server in Chicago. The system in Detroit can give the mobile device a list of (closer) redirect numbers to call. In this manner, a mobile device can choose a first and perhaps second alternate modem bank to call if numbers are busy, or if a primary choice cannot handle additional connections.

It is possible that a call may be originated by the network toward the mobile user. In order to facilitate this, a CS-CDPD mobile must first register with the network by placing a call and executing the procedures outlined above. Within the CM-ESs original CSCCP connection request, a dial code parameter exists. The dial code parameter is simply the number to call if this CM-ES needs to be reached. By providing this number in the original connection request, a serving CMD-IS knows how to reach a mobile within its serving area. Home MD-ISs know which server to forward calls to via the mobile data location procedures that exist between both MD-ISs.

In Fig. 9-16 we see that modem pools are co-located with CMD-ISs. This does not have to be the case. A modem pool and CMD-IS can be remotely located and connected together via wide area technologies such as frame relay, or even private lines. This was implied in Fig. 9-14. The hybrid will enable rapid deployment of network services, and will be an important step toward making service available to all users. This is true because a complete (CS-CDPD) system would not have to be deployed in each area of AMPS coverage; only a modem pool and access number would be required. Therefore, time and money will be saved by provisioning remote access capability within each service area and tying these consolidation points to remote CSMD-ISs. Deployment of a complete CDPD system within each service area, with MDBS capability at each cell site, is a time-consuming and expensive task. Circuit switched CDPD is a very attractive and much-needed short-term coverage solution. It will provide ubiquitous access capability to network services with less than ubiquitous coverage of a CDPD signal. (Refer to Fig. 9-17.)

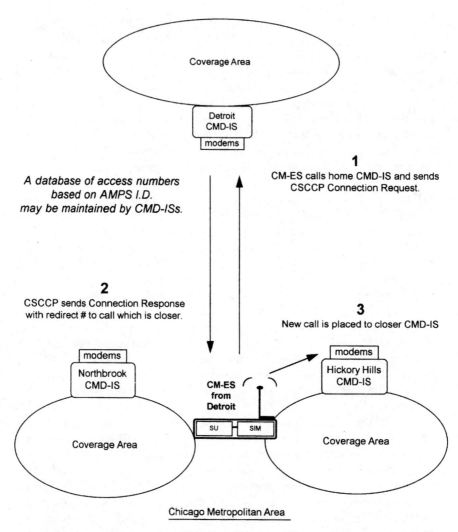

A database of access numbers based on AMPS I.D. may be maintained by CMD-ISs.

1
CM-ES calls home CMD-IS and sends CSCCP Connection Request.

2
CSCCP sends Connection Response with redirect # to call which is closer.

3
New call is placed to closer CMD-IS

Figure 9-16 CS-CDPD has the ability to redirect a calling party to a more suitable access number.

Closing

Some like to argue that certain technologies will "win" while others will "lose." We argue that no one technology will be "the winner." Instead, each tool that is available will complement other tools already in use. The only "winner" will be the community of users. Every application has its own needs, and users will always choose the network technology that "makes sense" for their unique needs.

But just for a moment, consider the size of the corporate population

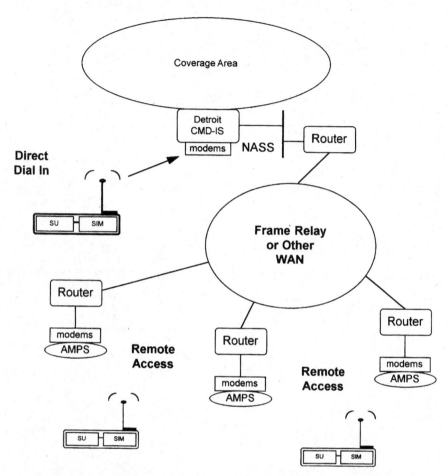

Figure 9-17 Dispersed modem pools which serve CS-CDPD users can be supplemented with third-party network services to provide wide area coverage of network capabilities.

already connected to the Internet, and the ever-increasing number of corporations going on-line each week. CDPD provides a natural way to communicate with these businesses in a mobile manner, and many corporations are discovering that it does work as advertised. The age of mobile computing is indeed upon us, so if your organization is considering a mobile communications tool, consider CDPD. It might just be what you have been looking for.

CDPD Assigned Numbers*

Identifier	Format	Assigned by	Notes
Routing area identifier	Number: 16 bits (0–65,535)	Service provider network	Unique within routing domain. Forms area field of OSI-CLNP NEI.
System identifier	Number: 6 octets	Same as above	Unique within routing domain. Forms system field of OSI NEI.
Internet network identifier (net ID)	As defined in class A,B,C IP formats	Internet, as delegated to CDPD NIC, then delegated to service provider network	Unique within Internet. Forms high-order part of Internet NEI.
Host identifier	Same as above	Service provider network	Unique within network. Forms low-order part of Internet NEI.
Subscriber identifier	Number: 32 bits	Same as above	Provided in remote activation.
RF channel identifier	10 bits (1–1023)	EIA/TIA 553	CDPD uses standard channel IDs.
Temporary equipment identifier (TEI)	Number: 27 bits (0–134,217,727)	Serving MD-IS	Assigned to M-ES in identity assign message. Not visible outside MDLP.
Customer identifier	Number: 32 bits	Service provider	Provided in remote activation.
Accounting distributor identifiers	Number: 16 bits	Same as above	Inserted in home accounting segment by serving accounting distributor.
Consolidation accounting collector identifier	Same as above	Same as above	Inserted in consolidation accounting segment by home accounting distributor.

(Continued)

*From CDPD R1.1, Introduction and Overview, Appendix 101A.

Identifier	Format	Assigned by	Notes
Preferred interexchange service provider code (PIC)	Number: 16 bits (1–65,535) Name: 64 characters	CDPD NIC	Carried as part of the HOMEINFO field.
Service group identifier	Number: 16 bits (1–65,535) Name: 64 characters	Service provider	Combination of SPI, SPNI, and/or WASI
Wide area service identifier (WASI)	Number: 16 bits (1–65,535) Name: 64 characters	CDPD NIC	Globally distinct. Broadcast in channel ID message, can be stored in M-ES, and part of traffic data matrix.
Accounting area identifier	Number: 16 bits (1–65,535)	Service provider	Georgraphic designation.
Service provider identifier (SPI)	Number: 16 bits (1–65,535) Name: 64 characters	CDPD NIC	Globally distinct, part of channel ID message. Can be stored by M-ES and is part of accounting matrix.
Service provider network identifier (SPNI)	Same as above	Same as above	Globally distinct. Broadcast in channel ID message as part of cell identifier and stored in M-ES.
Routing domain identifier	Number: 16 bits (0–65,535) Name: 64 characters	Same as above	Globally distinct. Forms RD field of OSI NEI.
Business relationship identifier (BRI)	Number: 16 bits (1–65,535)	CDPD NIC	Globally distinct. Included in HOMEINFO field and traffic data matrix.
Cell identifier—global (cell ID)	Number: 32 bits (1–65,535 for SPNI) (1–65,535 within SPNI)	Service provider network	Globally distinct. High-order 16 bits are SPNI. Included in traffic matrix data row.
Local service distributor identifier	Number: 16 bits (1–65,535)	Service provider	HLS-ID carried within HOMEINFO field.
Channel stream identifier (CSI)	Number: 6 bits (0–63)	Same as above	Unique within cell. Broadcast in channel ID message.
Color code—cell group	Number: 5 bits (0–31)	Same as above	Broadcast in each forward channel block.
Color code—area	Number: 3 bits (0–7)	Same as above	Same as above.
Multicast group member identifier (GMID)	Number: 16 bits (1–65,535)	Owner of multicast NEI	Used within multicast NEI. Sent by M-ES within ESH message and included in traffic data matrix.
Equipment identifier (EID)	Number: 48 bits	Equipment manufacturer, IEEE MAC address	Sent by M-ES in Identity Request and End System Hello messages.

B

TEI Management
Messages

TEI Identity Request
V.2

8	7	6	5	4	3	2	1

8	7	6	5	4	3	2	1
TEI = 0						0 CR	1 E
0	0	0	0	0	0	1	1
Control Field = UI							
LMEI = 15 (TEI Administration)							
Message Type = 1							
Equipment I.D. (4 bytes IEEE MAC)							
M-ES MDLP Version							
Transmit Window - K							
Receive Window - K							
M-ES Retransmission Timer - T200 2 bytes, increments of 1/10 second 10 min - 600 max - 30 recommended							
Retransmission Counter - N200 2 min - 10 max - 3 recommended							
Idle Timer - T203 2 bytes, increments of 1 second At least 2 x T200 - max 3600 - recommended 30							
Maximum TEI Notification Attempts - N204 1 min - 255 max - 5 recommended							
ACK Delay Timer - T205 Increments of 1/10 second 0 min - 255 max - 5 recommended Should be less than T200/2							
V.42 bis Data Compression Request (P0)							
V.42 bis Number of Codewords (P1)							
V.42 bis Maximum String Length (P2)							

TEI Identity Assign
V.2

8	7	6	5	4	3	2	1
			TEI = 0			1 CR	1 E
0	0	0	0	0	0	1	1
			Control Field = UI				
			LMEI = 15 (TEI Administration)				
			Message Type = 2				
			Equipment I.D. (4 bytes IEEE MAC)				
		ASSIGNED TEI VALUE Up to 4 bytes in length				Ext.	
			M-ES MDLP Version				
			Transmit Window - K				
			Receive Window - K				
		M-ES Retransmission Timer - T200 2 bytes, increments of 1/10 second 10 min - 600 max - 30 recommended					
		Retransmission Counter - N200 2 min - 10 max - 3 recommended					
		Idle Timer - T203 2 bytes, increments of 1 second At least 2 x T200 - max 3600 - recommended 30					
		Maximum TEI Notification Attempts - N204 1 min - 255 max - 5 recommended					
		ACK Delay Timer - T205 Increments of 1/10 second 0 min - 255 max - 5 recommended Should be less than T200/2					
		V.42 bis Data Compression Request (P0)					
		V.42 bis Number of Codewords (P1)					
		V.42 bis Maximum String Length (P2)					

TEI Identity Check Message
V.2

8	7	6	5	4	3	2	1
TEI = 0						1 CR	1 E
0	0	0	0	0	0	1	1
Control Field = UI							
LMEI = 15 (TEI Administration)							
Message Type = 4							
ASSIGNED TEI VALUE Up to 4 bytes in length						Ext.	

TEI Identity Check Response Message
V.2

8	7	6	5	4	3	2	1
TEI = 0						0 CR	1 E
0	0	0	0	0	0	1	1
Control Field = UI							
LMEI = 15 (TEI Administration)							
Message Type = 5							
4 bytes of EID							
ASSIGNED TEI VALUE Up to 4 bytes in length						Ext.	

TEI Identity Remove Message
V.2

8	7	6	5	4	3	2	1
TEI = 0						1 CR	1 E
0	0	0	0	0	0	1	1
Control Field = UI							
LMEI = 15 (TEI Administration)							
Message Type = 6							
ASSIGNED TEI VALUE Up to 4 bytes in length						Ext.	

TEI Notification Message
V.2

8	7	6	5	4	3	2	1
TEI = 0						1 CR	1 E
0	0	0	0	0	0	1	1
Control Field = UI							
LMEI = 1 (Sleep Mode Management)							
Message Type = 1 (TEI notification)							
TEI LIST All TEIs requiring notification						Ext.	

C

Information Sources
for the Reader

Home Pages and Internet Points of Reference

Each of these locations on the World Wide Web points to many other useful depositories of information.

Ameritech Home Page   http://www.ameritech.com
CDPD Forum Home Page   http://www.cdpd.org/
Communications and Telecommunications Virtual Library
   http://www.analysys.co.uk/commslib.htm
Internet RFCs   gopher://sunic.sunet.se:7532/11/
Mobile and Wireless Computing Home Page
   http://snapple.cs.washington.edu:600/mobile/mobile_www.html

Vendors in the CDPD Market

This list is not an all-inclusive list of CDPD vendors; it is merely a listing of some of the companies that are currently providing CDPD-related goods and services. Many other vendors who provide CDPD-related solutions do not appear on this list. No vendor has been intentionally omitted from this reference. Only sources known to us are listed.

Agosta and Associates
2217 Briar Hill Drive
Schaumburg, IL 60194
v (847) 686-1022
f (847) 686-1022
e-mail jagosta@interaccess.com

Hands-on workshops and data communications seminars, including CDPD.

AirLink Communications, Inc.
Contact: James Baichtal, Ray Fasnacht
v (408) 261-6602
f (408) 261-6680
e-mail ray@AirLink.com

AirLink Communications develops CDPD protocol software and builds CDPD test equipment that is used by CDPD equipment manufacturers and carriers.

Aldiscon, Inc.
201 Bradenton Ave.
Dublin, OH 43017
v (614) 764-2490
f (614) 764-2461

Short messaging services over CDPD.

Ameritech
2000 West Ameritech Center Dr.
Hoffman Estates, IL 60195-5000
v (708) 765-5834
f (708) 765-3702

CDPD network services.

Atronet Corporation
37 Skyline Drive
Lake Mary, FL 32746
v (407) 333-4926
f (407) 333-4966

MDBS equipment for campus environments.

ATT Network Wireless Systems
67 Whippany Road
Whippany, NJ 07040
v (201) 386-7765
f (201) 386-2383

CDPD systems and applications vendor.

ATT Wireless Services
10230 Northeast Points Drive
Kirkland, WA 98033
v (206) 803-4617
f (206) 803-4601

Provider of wireless voice and data communications services, including CDPD.

Bell Atlantic NYNEX Mobile

The largest wireless service company on the East Coast; provides customers with a full range of wireless voice, paging, and data communications solutions. For information, call (800) 308-DATA or, on the Internet, http://www.banm.com.

The Bishop Company
1125 East Milham Road
Kalamazoo, MI 49002
Dave Chamberlain
v (800) 520-7766
v (616) 381-9416
f (616) 381-9850

Educational services.

CDPD Forum, Inc.
401 N. Michigan Ave.
Chicago, IL 60611
v (312) 644-6610
f (312) 321-6869

An industry association that brings together companies which provide CDPD products and services. The forum's mission is to promote the technology and to develop the CDPD marketplace.

Center for Software Development
111 W. John Street
San Jose, CA 95113
v (408) 494-8303
f (408) 494-8383

Operates a self-service lab designed specifically for wireless testing in a mobile communications environment.

Ericsson, Inc.
740 East Campbell Road
Richardson, TX 75081
v (214) 437-8188
f (214) 705-7888

Provides a complete CDPD infrastructure, including R1.1.

First Net Corporation
9461 LBJ Freeway, Suite 100
Dallas, TX 75243
v (214) 231-6655
f (214) 231-6110

Point-of-sales services.

GTE Mobilnet
245 Perimeter Center Parkway
Atlanta, GA 30346
v (404) 395-8783
f (404) 395-8733

CDPD and CS-CDPD applications.

IBM
P.O. Box 12195
Durham, NC 27709
v (919) 543-6867
f (919) 543-7378

CDPD applications.

IFR Systems, Inc.
10200 West York Street
Wichita, KS 67215
v (316) 522-4981
f (316) 522-1360

Communications test equipment including CDPD capability.

Isotel Research Ltd.
6815 Eighth Street Northeast, Suite 360
Calgary, AB T2E 7H7
Canada
v (403) 275-0041
f (403) 274-3598

CDPD control systems gateway.

Marben Products, Inc.
2105 Hamilton Avenue
San Jose, CA 95125
v (408) 879-4001
f (408) 879-4001

OSI software products and integration services.

MobileWare Corporation
2425 N. Central Expressway No. 1001
Richardson, TX 75080
v (214) 952-1271
f (214) 690-6185

CDPD application software.

Motorola
50 East Commerce Drive
Schaumburg, IL 60173
v (847) 576-8923
f (847) 538-4591

CDPD wireless modems, PCMCIA and 3-watt mobiles.

MPR Teltech, Inc.
8999 Nelson Way
Burnaby, BC V5A 4B5
Canada
v (604) 293-6149
f (604) 293-5754

CDPD analysis tools.

NORTEL (Northern Telecom)
2221 Lakeside Blvd.
Richardson, TX 75082
v (214) 684-7968
f (214) 684-3943

Wireless data applications.

Packet Cluster Systems
2 Mount Royal Avenue
Marlborough, MA 01752
v (508) 460-4000
f (508) 460-4099

Mobile data communications for law enforcement and public safety.

PCSI
9645 Scranton Road
San Diego, CA 92121
v (619) 535-9500
f (619) 535-0106

CDPD mobile and infrastructure equipment.

Products Research, Inc.
1550 A Fullerton Ave.
Addison, IL 60101
v (847) 627-3337
f (847) 627-1126

Telemetry and GPS applications.

Racotek, Inc.
7301 Ohms Lane, No. 200
Minneapolis, MN 55439
v (612) 893-3940
f (612) 832-9383

Applications programming interface for CDPD access.

Radiomail Corporation
2600 Campus Drive
San Mateo, CA 94403
v (415) 286-7830
f (415) 286-7805

Messaging services including CDPD.

Retix
2401 Colorado Avenue
Santa Monica, CA 90404
v (310) 838-3400
f (310) 838-2255

CDPD accounting and network management systems.

Sierra Wireless Inc.
13151 Vanier Place
Richmond, BC V6V 2J2
Canada
v (604) 231-1106
f (604) 231-1109

Multimode wireless modems, including CDPD.

Software Corporation of America
100 Prospect Street
Stamford, CT 06901
v (203) 359-2773
f (203) 359-3198

Wireless dispatch and CDPD applications.

Sony Electronics, Inc.
16450 West Bernardo Drive
San Diego, CA 92127
v (619) 673-2962
f (619) 673-3232

Route guidance and fleet management.

Steinbrecher Corporation
30 North Avenue
Burlington, MA 01803
v (617) 273-1400
f (617) 273-4160
http://steinbrecher.com

Steinbrecher supplies wideband base station products for wireless communications applications, including cellular, CDPD, and PCS.

TEKnique, Inc.
911 North Plum Grove Road
Schaumburg, IL 60173
v (847) 706-9700
f (847) 706-9735

Client server applications for circuit switched and CDPD services.

Telecom Consultants Group
121 Pebble Drive
Clayton, NC 27520
v (919) 550-0802

Consulting services in the communications and telephone industry.

VLSI Technology
1109 Mckay Drive
San Jose, CA 95131
v (408) 434-3100
f (408) 922-5245

Small-form CDPD devices.

Wireless Connect, Inc.
2177 Augusta Place
Santa Clara, CA 95051
v (408) 296-1546
f (408) 448-3844

CDPD applications development tools.

WRQ
1500 Dexter Avenue North
Seattle, WA 98109
v (206) 217-7100
f (206) 217-0211

PC connectivity software, finetuned for mobile environments.

D

OSI Address Primer

The *connectionless network protocol* (CLNP) is an international standards-based IP "lookalike." Observation of the complete IP and CLNP headers shows that there are many fields within each protocol that serve identical purposes, albeit they are named differently. Figure D-1 shows the format of both the IP and CLNP headers. Similarities between CLNP and IP worth noting include the following.

CLNP	IP	Description
Version	Version	Protocol version being implemented.
Header length	Header length	Length of the header—useful if options are included.
Lifetime	Time to live	Maximum "hop count." this value is decremented by each router.
Total length	Total length	Total length of entire datagram.
Data unit I.D.	Identification	"Tag" identifying individual segments as belonging to a specific PDU.
SP MS	Flags	Identifies whether segmentation is allowed and whether more segments will follow.
Seg. offset	Frag. offset	Offset position of a segment/fragment with respect to entire PDU.
Options	Options	Options such as complete/partial, strict/loose source routing.

Obviously, there are differences worth noting as well. One difference is the "protocol" field found within IP, which does not have a peer function within CLNP. The protocol field identifies whether the higher-layer data carrier is UDP, TCP, ICMP, etc. The CLNP simply states

IP Connectionless Datagram Format

Bits	0 1 2 3	4 5 6 7	8 9 10 11 12 13 14 15	16 17 18 19 20 21 22 23 24 25 26 27 28 29 30 31
Octets 1 - 4	Version	Header Length	Service Type	Total Length
5 - 8	. Identification			Flags	Fragment Offset
9 - 12	Time to Live		Protocol	Header Checksum
13 - 16	Source IP Address			
17 - 20	Destination IP Address			
21 - 24	IP Options (if any)			Padding
Variable	DATA (TCP or other higher-layer protocols)			

ISO CLNP 8473 Header Format

8	7	6	5	4	3	2	1
Network Layer Protocol Identifier (81)							
Header Length Indicator							
Version							
Lifetime							
SP	MS	ER	Type				
Segment Length							
Header Checksum							
Destination Address Length							
Destination Address							
Source Address Length							
Source Address							
Data Unit Identifier							
Segment Offset							
Total Length							
Options (may or may not be present)							
DATA							

Figure D-1

whether the "type" of PDU is a data PDU or an error report PDU. CLNP error report PDUs are analogous to the Internet Control Message Protocol (ICMP) in TCP/IP implementations. The CLNP uses a single-byte field within the address portion to indicate what type of data or upper-layer protocols is being serviced. This field is called an *NSAP selector*.

One key difference between IP and CLNP is the address fields. While IP addresses are fixed in length at 4 bytes, CLNP addresses can vary in length. This appendix provides a brief tutorial on CLNP address structures.

One of the goals of CLNP addressing is to accommodate existing address plans rather than specifying a new, global communications address plan. If there was only one global authority in charge of a single network domain, there would be very little structure reflecting network types and locations. Incredibly huge directory systems would be required for name, address, and routing information. As we know, this is not the case; there are many different types of networks, both public and private. Because of this, CLNP is capable of identifying the "type of address" and "numbering plan information" (TOA/NPI) being used within each datagram, for each network type. Once identification of address plan information has been accomplished, the address itself may occupy the corresponding fields within CLNP's header. A CLNP address is also known as an NSAP, or *network service access point address*. Let's look at the generic format of the CLNP address fields, illustrated in Fig. D-2.

As the top of the figure illustrates, the maximum length of a CLNP address is 20 bytes. However, depending on the environment in which CLNP is being used, the address itself may be less than 20 bytes in

OSI NSAP Address Format - Generic

	IDP		DSP
	AFI	IDI	Defined by Authority Identified by IDP
20 bytes max.	1	Variable	20 - (IDI +AFI)

OSI NSAP Address Format - United States Government

	IDP		DSP	
	AFI	IDI	NSAP Addresses	SEL
Bytes	1	2	16	1

Figure D-2

length. There are two main parts to an NSAP address: the initial domain part (IDP) and the domain-specific part (DSP). The IDP portion is divided into two subfields. These subfields are known as the *authority and format identifier* (AFI) and the *initial domain identifier* (IDI).

Authority Format Identifier

The AFI is used to identify "the format of the IDI, the authority responsible for allocating IDI values, and the syntax of the DSP." IDI authorities have been defined in ISO/IEC 8348 as indicated in the following list. All values are indicated in decimal.

AFI	Authority
00–09	Reserved and will not be allocated.
10–35	Reserved for future use by joint CCITT and ISO/IEC.
36–59	Allocated to reflect IDI formats as indicated in Fig. D-3 (Table A.2 of international standard).
60–69	Allocated for ISO/IEC-specific IDI formats.
70–79	Allocated by CCITT/ITU for new IDI formats.
80–99	Reserved for future allocation by joint agreement of ISO and CCITT.

Most AFI values used, including U.S. government and industry consortiums such as the CDPD consortium, fall into the range 36–59. The bottom of Fig. D-2 shows how the IDP and DSP are further divided for U.S. addressing. Of the total length of 20 bytes, 1 byte within the DSP is used for the higher-layer protocol identifier, or NSAP selector, while the first 3 bytes occupy the IDP portion. Accordingly, only 16 bytes of information are related directly to the network-level entity source and destinations. Figure D-3 shows how each AFI value in the range 36–59 is used.

Initial Domain Indicator

The two AFI values 47 and 39 are commonly used. As Fig. D-3 indicates, an AFI of 47 implies that the syntax of the domain-specific part is binary notation. However, the actual authority governing the dissemination of the DSP is defined within the IDI part of the address. As assigned by the ISO in 6523, the U.S. government has been assigned four IDI values. U.S. government IDI values have been assigned for the DOD, OSINET, GOSIP, and OSI Implementors Workshop domains. For example, the IDI value 0005 is owned by the US GOSIP (U.S. Government Open Systems Interconnect Profile) authority, which is the General Services Administration's information resource

Allocated AFI Values and IDI Formats

IDI FORMAT	DSP SYNTAX			
	Decimal	Binary	ISO/IEC 646 Character	National Character
X.121	36, 52	37, 53	na	na
ISO DCC	38	39	na	na
F.69	40, 54	41, 55	na	na
E.163	42, 56	43, 57	na	na
E.164	44, 58	45, 59	na	na
ISO 6523	46	47	na	na
Local	48	49	50	51
Terminology	**Description**			
Decimal	Expect symbols in the range of 0 - 9 (nibbles 0000 - 1001)			
Binary	Expect symbols in the range 0 - F (nibbles 0000 - 1111)			
ISO Character	A character in the range of 0 - 95 as per ISO 646			
National Char.	Within the United States, IA5 or ASCII expressed in 8 bits			
X.121	Public Data Networks such as X.25 numbering			
ISO DCC	ISO Data Country Code / 840 United States			
F.69	International Telex numbering plan			
E.163	Telephony - 4-digit PSTN code and 12-digit network number			
E.164	ISDN			
ISO 6523	Specifies a 4-digit International Code Designator as per ISO 6523			
Local	Non - OSI addresses, user-specific			

Figure D-3

management service (IRMS). Therefore, the IDP (decimal) combination 47/0005 is the AFI/IDI of a US GOSIP address. Notice that the IDI portion is 2 bytes in length, with each half-octet representing one of the four digits making up the ICD code assigned by ISO 6523.

Not all IDI indicators, however, are 4-byte codes occupying 2 octets. Depending on the AFI indicator, the IDI part can be constructed in any of seven ways. Again, Fig. D-3 illustrates this point. What follows is a summary of the various IDI formats that have been defined.

X.121   An AFI of 2 digits plus an IDI of up to 14 digits, for a total IDP length of 16 digits, or 8 octets. The full X.121 number indicated in the IDP identifies the authority responsible for allocation of the remaining octets found in the DSP portion.

ISO DCC   A 2-digit AFI and a 3-digit country code as allocated by ISO 3166. The code 840 is assigned to the United States. Total IDP length is 5 digits represented in 3 bytes. Again, the IDP identifies the authority for assigning the DSP. The American National Standards Institute (ANSI) is the appropriate body under the USA DCC code, 840.

F.69   Total IDP length is up to 10 digits. IDP consists of a 2-digit AFI followed by a telex number in the IDI portion. A telex number is up to 8 digits in length and precedes with a 2- or 3-digit destination "area code." For example, the digits "23" are a 2-digit F-69 area code for the United States, while the digits "851" comprise the 3-digit F-69 area code for the United Kingdom. Again, the telex number indicated within the IDP only identifies the authority in control of the DSP portion of the address.

E.163   Consists of a PSTN number of up to 12 digits in length, beginning with the E.163 country code. The number identifies the authority responsible for the DSP of the address. Total IDP length is up to 14 digits, including the AFI.

E.164   The IDI consists of an ISDN number of up to 15 digits in length, commencing with the country code defined in E.164. Total IDP length is up to 17 digits, including the AFI.

ISO 6523 ICD   A 4-digit country code assigned by ISO 6523. 0005 and 0006 belong to the U.S. government. Total IDP length is 6 digits, including the AFI.

Local IDI   The IDI portion is not used. After the AFI indicates "local" format, the remaining information is the DSP itself, coded according to national (USA = ANSI) or ISO/IEC 646 rules.

U.S. IDP Formats

Examples are offered to illustrate various NSAP address schemes used within the United States. The first example, a US GOSIP NSAP address, is shown in Fig. D-4. The second example, using the DCC format, is shown in Fig. D-5. The central authority in the United States for DCC-encoded NSAP addresses is ANSI.

CDPD NSAP Address Formats

CDPD falls under the authority of ANSI. Therefore, CDPD NSAP addresses use the DCC format and an AFI/IDI of 39/840 decimal. Figure D-6 identifies the CDPD-specific NSAP structure, while Fig. D-7 identifies the NSAP selectors used with the CDPD service. The DFI value 128 (80 hex) indicates that the ensuing portions of the DSP will follow GOSIP v.2 formats. GOSIP v.2 is part of the Industry/Government Open Systems Specification profile, which defines allowable protocols and procedures to use for communications purposes.

OSI NSAP Address Format - U.S. GOSIP

IDP		DSP						
AFI	IDI	DFI	AA	RES	RD	AREA	ID	SEL
Decimal 47	5	128	GSA	0	Defined by Agency Such as Dept. Interior, Dept. Commerce, Etc.			Variable
Hex 2F	0005	80	GSA	0000				Variable
Bytes 1	2	1	3	2	2	2	6	1

DFI   Domain format identifier. A value of 128 indicates GOSIP V.2.

AA   Administrative authority. These values are assigned by GSA.

RES   Reserved.

RD   Routing domain. Assigned by GSA / agency.

AREA   Area within routing domain. Assigned by agency.

ID   System identifier which is assigned by agency.

SEL   NSAP selector code for higher-level protocol such as TP4.

Figure D-4

OSI NSAP Address Format - USA DCC Format

IDP		DSP						
AFI	IDI	DFI	ORG	RES	RD	AREA	ID	SEL
Decimal 39	840	128	ANSI	0	These fields defined by ANSI organization			variable
Hex 27	0348	80	ANSI	0000				variable
Bytes 1	2	1	3	2	2	2	6	1

AFI   Value of 39 decimal calls for Data Country Code format.

IDI   Value of 840 decimal has been assigned to ANSI for USA domains.

DFI   Domain format identifier. A value of 128 indicates GOSIP (IGOSS) V.2.

ORG   Assigned by ANSI to various domains. CDPD has an assignment.

RES   Reserved.

RD   Routing domain. Assigned by organization.

AREA   Area within routing domain. Assigned by organization.

ID   System identifier which is assigned by organization.

SEL   NSAP selector code for higher-level protocol such as TP4.

Figure D-5

OSI NSAP Address Format - CDPD

	IDP		DSP						
	AFI	IDI	DFI	ORG	RES	RD	AREA	ID	SEL
Decimal	39	840	128	113660	0	These fields defined by CDPD carrier. See Appendix A.			0 - 5
Hex	27	0348	80	01BBFC	0000				00 - 05
Bytes	1	2	1	3	2	2	2	6	1

AFI   Value of 39 decimal calls for Data Country Code format.

IDI   Value of 840 decimal has been assigned to ANSI for USA domains.

DFI   Domain format identifier. A value of 128 indicates GOSIP (IGOSS) V.2.

ORG   Assigned by ANSI to CDPD domain.

RES   Reserved.

RD   Routing domain. Assigned by CDPD carrier.

AREA   Area within routing domain. Assigned by CDPD carrier.

ID   System identifier which is assigned by CDPD carrier.

SEL   NSAP selector code for higher-level protocol such as MNLP.

Figure D-6

CDPD
NSAP Selector Codes

0	CLNP
1	TP4
2	MDBS RRM
3	MNLP Location Updates
4	MNLP Forwarding
5	Channel Status

Figure D-7

Mobile Network Registration and Location Protocols

This appendix provides the reader with a complete listing of all MNRP and MNLP PDUs, and identifies message formats, result codes, parameters, and options associated with these protocols.

M - ES TEI	INFO	NLPI - 0	ESH, ESB, ISC, ESQ
Address	Control	SNDCP	MNRP PDUs and Options

MNRP
Message Format

Mobile Network Registration Protocol

RFC791 IP addresses are presented to MD - IS within the CLNP address structure. This applies only to the registration process.

Format of MNRP PDUs

PDU Type Identifier
Address Length - 7 bytes
OSI Address AFI = 49
IDI = 00
DFI = 00
4 bytes of IP Address
MNRP Options

OSI Format for RFC791 IP Addresses

8	7	6	5	4	3	2	1
Length Indicator - 7 bytes							
AFI - 49							
IDI - 00							
DFI - 00							
4 bytes of RFC791 IP addressing							

MNRP End System Hello - ESH

8	7	6	5	4	3	2	1
01 - ESH PDU Type							
Source Address Length							
Source Network Address / NEI							
Registration Counter							
Options - 0, 2							

MNRP End System Bye - ESB

8	7	6	5	4	3	2	1
02 - ESB PDU Type							
Source Address Length							
Source Network Address / NEI							
Options - 0							

MNRP MD - IS Hello Confirm (ISC)

8	7	6	5	4	3	2	1
03 - ISC PDU Type							
Destination Address Length							
Destination Network Address / NEI							
Options - 03, 04, 05							

MNRP MD - IS End System Query (ESQ)

8	7	6	5	4	3	2	1
04 - ESQ PDU Type							
Destination Address Length							
Destination Network Address / NEI							

MNRP Options

·8	7	6	5	4	3	2	1
Option Identifier								
Length of Parameters								
Parameter Values								
Next Option Identifier								
Length of Parameters for (second) Option								
Parameter Values for (second) Option								
Additional Option I.D.s, Length Indicators, and Parameters as appropriate......								

MNRP Option Value	MNRP Option Name	Size	Comments
0	Group Member Identifier	2 bytes	Unique among members of a multicast group.
1	EID - Equipment Identifier (obsolete)	4 bytes	IEEE MAC address.
2	Authentication Information	11 bytes	Algorithm I.D., ASN, ARN.
3	Authentication Update	9 bytes	Algorithm I.D., ARN.
4	Result Code	1 byte	Result code I.D.
5	Configuration Timer	2 bytes	Indicated in seconds.

MNRP / MNLP Authentication Option

8	7	6	5	4	3	2	1
Option Code - 2							
Length Identifier - 11							
Authentication Algorithm Identifier - 0 (CDPD ver 1)							
2 bytes ASN							
8 bytes ARN							

MNRP / MNLP Authentication Update Option

8	7	6	5	4	3	2	1
Option Code - 3							
Length Identifier - 9							
Authentication Algorithm Identifier - 0 (CDPD ver 1)							
8 bytes ARN							

MNRP Result Codes

Value	Meaning
0	Registration accepted
1	No particular reason given
2	MD - IS not capable of registering M - ES at this time
3	M - ES not authorized to use this subnetwork
4	M - ES gave insufficient authentication credentials
5	M - ES gave unsupported authentication credentials
6	NEI has exceeded usage limitations
7	Service denied on this subnetwork; service may be obtained on alternate service provider network
8 - 255	Reserved

Mobile Network Location Protocol Message Identifiers

PDU Type	Interpretation
1	Redirect Request (RDR)
2	Redirect Confirm (RDC)
3	Redirect Flush (RDF)
4	Redirect Expiry (RDE)
5	Redirect Query (RDQ)

MNLP Redirect Request (RDR)

8	7	6	5	4	3	2	1
RDR PDU - Type 1							
Registration Sequence Count							
Source Address Length							
Source Address							
Forwarding Address Length							
Forwarding Address							
Options - 0, 2, 9							

MNLP Redirect Confirmation (RDC)

8	7	6	5	4	3	2	1
Redirect Confirmation PDU Identifier - 2							
Registration Sequence Count							
Destination Address Length							
Destination Address							
Options - 0, 3, 4, 5, 6, 8							

MNLP Redirect Flush (RDF)

8	7	6	5	4	3	2	1
Redirect Flush PDU Identifier - 3							
Registration Sequence Count							
Source Address Length							
Source Address							
Options - 0							

MNLP Redirect Expiry (RDE)

8	7	6	5	4	3	2	1
Redirect Expiry PDU Identifier - 4							
Registration Sequence Count							
Source Address Length							
Source Address							
Options - 0							

MNLP Redirect Query (RDQ)

8	7	6	5	4	3	2	1
Redirect Query PDU Identifier - 5							
Registration Sequence Count							
Destination Address Length							
Destination Address							

MNLP Options

8	7	6	5	4	3	2	1
Option Identifier								
Length of Parameters								
Parameter Values								
Next Option Identifier								
Length of Parameters for (second) Option								
Parameter Values for (second) Option								
Additional Option I.D.s, Length Indicators, and Parameters as appropriate......								

MNLP Option Value	MNLP Option Name	Size	Comments
0	Group Member Identifier	2 bytes	Unique among members of a multicast group.
1	EID - Equipment Identifier (obsolete)	4 bytes	IEEE MAC address.
2	Authentication Information	11 bytes	Algorithm I.D., ASN, ARN.
3	Authentication Update	9 bytes	Algorithm I.D., ARN.
4	Result Code	1 byte	Result code I.D.
5	Configuration Timer	2 bytes	Indicated in seconds.
6	Address Mask	variable	Traditional IP format.
7	Home Tariff Code (obsolete)	14 max	Was used for accounting.
8	Home Info Code	18 bytes	Provides treatment information to serving MD-IS such as preferred IEC to use as per home MD-IS.
9	Location Info	9 bytes	WASI, SPI, SPNI of server.
10 - 255	Reserved		

MNLP Result Code	MNLP Result Code Description
0	Registration accepted
1	No particular reason given
2	MD-IS cannot serve M - ES at moment
3	M - ES not authorized on this subnet
4	M - ES gave insufficient authentication credentials
5	M - ES gave unsupported authentication credentials
6	NEI has exceeded usage limitations
7	Service denied on this subnetwork but may be obtained on alternate carrier's network
8 - 255	Reserved

F

Data Traces

WORLD WIDE WEB SESSION TRACES
TRACE 1

This was the IP, ICMP, UDP, and DNS traffic generated as a result
of manually clicking on the following Mosaic URLs:

http://frame-relay.indiana.edu/
http://frame-relay.indiana.edu/frame-relay/5000/5000index.html
http://frame-relay.indiana.edu/frame-relay/5000/5001.html
http://frame-relay.indiana.edu/frame-relay/5000/5001-approved.html

```
ID:1536   0x0001    frame.relay.indiana.edu
UDP 32768 -> 53 SIZE : 41
00 06 01 00 00 01 00 00 00 00 00 00 0b 66 72 61    .............fra
6d 65 2d 72 65 6c 61 79 07 69 6e 64 69 61 6e 61    me-relay.indiana
03 65 64 75 00 00 01 00 01                         .edu.....
IP 149.174.213.15 -> 204.148.87.155, Length 40, Protocol 6
TCP 4214 <- 80 seq:05D7F9BD ACK  wnd:8191 ack:000001CC size:0
IP 147.225.1.2 -> 204.148.87.155, Length 241, Protocol 17
UDP 32768 <- 53 SIZE : 213
00 06 85 80 00 01 00 02 00 03 00 03 0b 66 72 61    .............fra
6d 65 2d 72 65 6c 61 79 07 69 6e 64 69 61 6e 61    me-relay.indiana
03 65 64 75 00 00 01 00 01 c0 0c 00 05 00 01 00    .edu............
00 c4 e0 00 17 05 73 74 6f 6e 65 03 75 63 73 07    ......stone.ucs.
69 6e 64 69 61 6e 61 03 65 64 75 00 c0 35 00 01    indiana.edu..5..
00 01 00 00 c4 e0 00 04 81 4f 0c 01 c0 3b 00 02    .........O...;..
00 01 00 00 c4 e0 00 09 06 69 75 67 61 74 65 c0    .........iugate.
3b c0 3b 00 02 00 01 00 00 c4 e0 00 0b 05 6d 6f    ;.;...........mo
6f 73 65 02 63 73 c0 3f c0 3b 00 02 00 01 00 00    ose.cs.?.;......
c4 e0 00 11 05 61 72 67 75 73 03 63 73 6f 04 75    .....argus.cso.u
69 75 63 c0 47 c0 68 00 01 00 01 00 00 c4 e0 00    iuc.G.h.........
04 81 4f 01 09 c0 7d 00 01 00 01 00 00 c4 e0 00    ..O...}........
04 81 4f fe bf c0 94 00 01 00 01 00 03 f4 80 00    ..O.............
04 80 ae 05 3a                  ....:
ID:1536   0x8085   frame.relay.indiana.edu   CNAME    1 0xc4e00017 stone.ucs.indiana.edu
ID:1536   0x8085   frame.relay.indiana.edu   ANS      1 0xc4e00004 129.79.12.1
TCP 3433 -> 80 seq:00000000 SYN  wnd:4096 options:04 02 B4 05 20 2F 54 48 size:4
IP 129.79.12.1 -> 204.148.87.155, Length 40, Protocol 6
TCP 3433 <- 80 seq:5FD43A00 SYN ACK  wnd:4096 ack:00000001 size:0
TCP 3433 -> 80 seq:00000001 ACK PSH  wnd:4096 ack:5FD43A01 size:458
IP 129.79.12.1 -> 204.148.87.155, Length 40, Protocol 6
TCP 3433 <- 80 seq:5FD43A01 ACK  wnd:4096 ack:000001CB size:0
TCP 3433 -> 80 seq:000001CB ACK  wnd:4096 ack:5FD43A01 size:0
IP 129.79.12.1 -> 204.148.87.155, Length 216, Protocol 6
TCP 3433 <- 80 seq:5FD43A01 ACK PSH  wnd:4096 ack:000001CB size:176
TCP 3433 -> 80 seq:000001CB ACK  wnd:4096 ack:5FD43AB1 size:0
IP 129.79.12.1 -> 204.148.87.155, Length 552, Protocol 6
```

```
TCP 3433 <- 80 seq:5FD43AB1 ACK  wnd:4096 ack:000001CB size:512
TCP 3433 -> 80 seq:000001CB ACK  wnd:4096 ack:5FD43CB1 size:0
IP 129.79.12.1 -> 204.148.87.155, Length 552, Protocol 6
TCP 3433 <- 80 seq:5FD43CB1 ACK  wnd:4096 ack:000001CB size:512
TCP 3433 -> 80 seq:000001CB ACK  wnd:3584 ack:5FD43EB1 size:0
IP 129.79.12.1 -> 204.148.87.155, Length 552, Protocol 6
TCP 3433 <- 80 seq:5FD43EB1 ACK  wnd:4096 ack:000001CB size:512
IP 129.79.12.1 -> 204.148.87.155, Length 104, Protocol 6
TCP 3433 <- 80 seq:5FD440B1 FIN ACK PSH wnd:4096 ack:000001CB size:64
TCP 3433 -> 80 seq:000001CB ACK  wnd:4096 ack:5FD440F2 size:0
TCP 3433 -> 80 seq:000001CB FIN ACK wnd:4096 ack:5FD440F2 size:0
IP 129.79.12.1 -> 204.148.87.155, Length 552, Protocol 6
TCP 3433 <- 80 seq:5FD43EB1 ACK  wnd:4096 ack:000001CB size:512
TCP 3005 -> 80 seq:00000000 SYN wnd:4096 options:04 02 B4 05 00 00 00 00 size:4
IP 129.79.12.1 -> 204.148.87.155, Length 40, Protocol 6
TCP 3433 <- 80 seq:5FD440F2 ACK  wnd:4095 ack:000001CC size:0
IP 129.79.12.1 -> 204.148.87.155, Length 40, Protocol 6
TCP 3005 <- 80 seq:5FE0EC00 SYN ACK  wnd:4096 ack:00000001 size:0
TCP 3005 -> 80 seq:00000001 ACK PSH wnd:4096 ack:5FE0EC01 size:529
IP 129.79.12.1 -> 204.148.87.155, Length 216, Protocol 6
TCP 3005 <- 80 seq:5FE0EC01 ACK PSH wnd:4096 ack:00000212 size:176
TCP 3005 -> 80 seq:00000212 ACK  wnd:4096 ack:5FE0ECB1 size:0
IP 129.79.12.1 -> 204.148.87.155, Length 552, Protocol 6
TCP 3005 <- 80 seq:5FE0ECB1 ACK  wnd:4096 ack:00000212 size:512
TCP 3005 -> 80 seq:00000212 ACK  wnd:4096 ack:5FE0EEB1 size:0
IP 129.79.12.1 -> 204.148.87.155, Length 552, Protocol 6
TCP 3005 <- 80 seq:5FE0EEB1 ACK  wnd:4096 ack:00000212 size:512
TCP 3005 -> 80 seq:00000212 ACK  wnd:4096 ack:5FE0F0B1 size:0
IP 129.79.12.1 -> 204.148.87.155, Length 552, Protocol 6
TCP 3005 <- 80 seq:5FE0F0B1 ACK  wnd:4096 ack:00000212 size:512
TCP 3005 -> 80 seq:00000212 ACK  wnd:4096 ack:5FE0F2B1 size:0
IP 129.79.12.1 -> 204.148.87.155, Length 552, Protocol 6
TCP 3005 <- 80 seq:5FE0F2B1 ACK  wnd:4096 ack:00000212 size:512
TCP 3005 -> 80 seq:00000212 ACK  wnd:4096 ack:5FE0F4B1 size:0
IP 129.79.12.1 -> 204.148.87.155, Length 552, Protocol 6
TCP 3005 <- 80 seq:5FE0F4B1 ACK  wnd:4096 ack:00000212 size:512
IP 129.79.12.1 -> 204.148.87.155, Length 41, Protocol 6
TCP 3005 <- 80 seq:5FE0F6B1 FIN ACK PSH wnd:4096 ack:00000212 size:1
TCP 3005 -> 80 seq:00000212 ACK  wnd:4096 ack:5FE0F6B3 size:0
TCP 3005 -> 80 seq:00000212 FIN ACK wnd:4096 ack:5FE0F6B3 size:0
IP 129.79.12.1 -> 204.148.87.155, Length 40, Protocol 6
TCP 3005 <- 80 seq:5FE0F6B3 ACK  wnd:4095 ack:00000213 size:0
TCP 1681 -> 80 seq:00000000 SYN wnd:4096 options:04 02 B4 05 20 2F 54 48 size:4
IP 129.79.12.1 -> 204.148.87.155, Length 40, Protocol 6
TCP 1681 <- 80 seq:5FFF3200 SYN ACK  wnd:4096 ack:00000001 size:0
TCP 1681 -> 80 seq:00000001 ACK PSH wnd:4096 ack:5FFF3201 size:531
IP 129.79.12.1 -> 204.148.87.155, Length 215, Protocol 6
TCP 1681 <- 80 seq:5FFF3201 ACK PSH wnd:4096 ack:00000214 size:175
TCP 1681 -> 80 seq:00000214 ACK  wnd:3921 ack:5FFF32B0 size:0
IP 129.79.12.1 -> 204.148.87.155, Length 543, Protocol 6
TCP 1681 <- 80 seq:5FFF32B0 FIN ACK PSH wnd:4096 ack:00000214 size:503
TCP 1681 -> 80 seq:00000214 ACK  wnd:4096 ack:5FFF34A8 size:0
TCP 1681 -> 80 seq:00000214 FIN ACK wnd:4096 ack:5FFF34A8 size:0
IP 129.79.12.1 -> 204.148.87.155, Length 40, Protocol 6
```

TCP 1681 <- 80 seq:5FFF34A8 ACK wnd:4095 ack:00000215 size:0
TCP 1714 -> 80 seq:00000000 SYN wnd:4096 options:04 02 B4 05 66 2F 61 72 size:4
IP 129.79.12.1 -> 204.148.87.155, Length 40, Protocol 6
TCP 1714 <- 80 seq:6009F000 SYN ACK wnd:4096 ack:00000001 size:0
TCP 1714 -> 80 seq:00000001 ACK PSH wnd:4096 ack:6009F001 size:557
IP 129.79.12.1 -> 204.148.87.155, Length 216, Protocol 6
TCP 1714 <- 80 seq:6009F001 ACK PSH wnd:4096 ack:0000022E size:176
TCP 1714 -> 80 seq:0000022E ACK wnd:3920 ack:6009F0B1 size:0
IP 129.79.12.1 -> 204.148.87.155, Length 552, Protocol 6
TCP 1714 <- 80 seq:6009F0B1 ACK wnd:4096 ack:0000022E size:512
TCP 1714 -> 80 seq:0000022E ACK wnd:4096 ack:6009F2B1 size:0
IP 129.79.12.1 -> 204.148.87.155, Length 552, Protocol 6
TCP 1714 <- 80 seq:6009F2B1 ACK wnd:4096 ack:0000022E size:512
IP 129.79.12.1 -> 204.148.87.155, Length 59, Protocol 6
TCP 1714 <- 80 seq:6009F4B1 FIN ACK PSH wnd:4096 ack:0000022E size:19
TCP 1714 -> 80 seq:0000022E ACK wnd:3565 ack:6009F4C5 size:0
TCP 1714 -> 80 seq:0000022E FIN ACK wnd:4096 ack:6009F4C5 size:0
IP 129.79.12.1 -> 204.148.87.155, Length 40, Protocol 6
TCP 1714 <- 80 seq:6009F4C5 ACK wnd:4095 ack:0000022F size:0
TCP 1767 -> 80 seq:00000000 SYN wnd:4096 options:04 02 B4 05 66 2F 61 72 size:4
IP 129.79.12.1 -> 204.148.87.155, Length 40, Protocol 6
TCP 1767 <- 80 seq:6014AE00 SYN ACK wnd:4096 ack:00000001 size:0
TCP 1767 -> 80 seq:00000001 ACK PSH wnd:4096 ack:6014AE01 size:561
TCP 1767 -> 80 seq:00000001 ACK PSH wnd:4096 ack:6014AE01 size:561
IP 129.79.12.1 -> 204.148.87.155, Length 216, Protocol 6
TCP 1767 <- 80 seq:6014AE01 ACK PSH wnd:4096 ack:00000232 size:176
TCP 1767 -> 80 seq:00000232 ACK wnd:4096 ack:6014AEB1 size:0
IP 129.79.12.1 -> 204.148.87.155, Length 552, Protocol 6
TCP 1767 <- 80 seq:6014AEB1 ACK wnd:4096 ack:00000232 size:512
TCP 1767 -> 80 seq:00000232 ACK wnd:4096 ack:6014B0B1 size:0
IP 129.79.12.1 -> 204.148.87.155, Length 552, Protocol 6
TCP 1767 <- 80 seq:6014B0B1 ACK wnd:4096 ack:00000232 size:512
TCP 1767 -> 80 seq:00000232 ACK wnd:3584 ack:6014B2B1 size:0
IP 129.79.12.1 -> 204.148.87.155, Length 552, Protocol 6
TCP 1767 <- 80 seq:6014B2B1 ACK wnd:4096 ack:00000232 size:512
TCP 1767 -> 80 seq:00000232 ACK wnd:4096 ack:6014B4B1 size:0
IP 129.79.12.1 -> 204.148.87.155, Length 552, Protocol 6
TCP 1767 <- 80 seq:6014B4B1 ACK wnd:4096 ack:00000232 size:512
TCP 1767 -> 80 seq:00000232 ACK wnd:3584 ack:6014B6B1 size:0
IP 129.79.12.1 -> 204.148.87.155, Length 552, Protocol 6
TCP 1767 <- 80 seq:6014B6B1 ACK wnd:4096 ack:00000232 size:512
TCP 1767 -> 80 seq:00000232 ACK wnd:3584 ack:6014B8B1 size:0
IP 129.79.12.1 -> 204.148.87.155, Length 552, Protocol 6
TCP 1767 <- 80 seq:6014B8B1 ACK wnd:4096 ack:00000232 size:512
IP 129.79.12.1 -> 204.148.87.155, Length 552, Protocol 6
TCP 1767 <- 80 seq:6014BAB1 ACK wnd:4096 ack:00000232 size:512
IP 129.79.12.1 -> 204.148.87.155, Length 552, Protocol 6
TCP 1767 <- 80 seq:6014BCB1 ACK wnd:4096 ack:00000232 size:512
IP 129.79.12.1 -> 204.148.87.155, Length 552, Protocol 6
TCP 1767 <- 80 seq:6014BEB1 ACK wnd:4096 ack:00000232 size:512
IP 129.79.12.1 -> 204.148.87.155, Length 552, Protocol 6
TCP 1767 <- 80 seq:6014C0B1 ACK PSH wnd:4096 ack:00000232 size:512
IP 129.79.12.1 -> 204.148.87.155, Length 552, Protocol 6
TCP 1767 <- 80 seq:6014C2B1 ACK wnd:4096 ack:00000232 size:512

TCP 1767 -> 80 seq:00000232 ACK wnd:3584 ack:6014C4B1 size:0
IP 129.79.12.1 -> 204.148.87.155, Length 552, Protocol 6
TCP 1767 <- 80 seq:6014C4B1 ACK PSH wnd:4096 ack:00000232 size:512
IP 129.79.12.1 -> 204.148.87.155, Length 552, Protocol 6
TCP 1767 <- 80 seq:6014C6B1 ACK wnd:4096 ack:00000232 size:512
TCP 1767 -> 80 seq:00000232 ACK wnd:3072 ack:6014C8B1 size:0
IP 129.79.12.1 -> 204.148.87.155, Length 552, Protocol 6
TCP 1767 <- 80 seq:6014C8B1 ACK wnd:4096 ack:00000232 size:512
TCP 1767 -> 80 seq:00000232 ACK wnd:3584 ack:6014CAB1 size:0
IP 129.79.12.1 -> 204.148.87.155, Length 169, Protocol 6
TCP 1767 <- 80 seq:6014CAB1 FIN ACK PSH wnd:4096 ack:00000232 size:129
TCP 1767 -> 80 seq:00000232 ACK wnd:4096 ack:6014CB33 size:0
TCP 1767 -> 80 seq:00000232 FIN ACK wnd:4096 ack:6014CB33 size:0
IP 129.79.12.1 -> 204.148.87.155, Length 169, Protocol 6
TCP 1767 <- 80 seq:6014CAB1 FIN ACK PSH wnd:4096 ack:00000232 size:129
TCP 1767 -> 80 seq:00000232 FIN ACK wnd:4096 ack:6014CB33 size:0
IP 129.79.12.1 -> 204.148.87.155, Length 40, Protocol 6
TCP 1767 <- 80 seq:6014CB33 ACK wnd:4095 ack:00000233 size:0

WORLD WIDE WEB SESSION TRACES
TRACE 2

This was the IP, ICMP, UDP, and DNS traffic generated by
clicking on the Mosaic URL:

http://frame-relay.indiana.edu/frame-relay/5000/5001-approved.html

ID:768   0x0001   frame.relay.indiana.edu
UDP 32768 -> 53 SIZE : 41
00 03 01 00 00 01 00 00 00 00 00 00 0b 66 72 61   ....fra
6d 65 2d 72 65 6c 61 79 07 69 6e 64 69 61 6e 61   me-relay.indiana
03 65 64 75 00 00 01 00 01   .edu.....
IP 198.80.0.6 -> 204.148.87.155, Length 241, Protocol 17
UDP 32768 <- 53 SIZE : 213
00 03 81 80 00 01 00 02 00 03 00 03 0b 66 72 61   ....fra
6d 65 2d 72 65 6c 61 79 07 69 6e 64 69 61 6e 61   me-relay.indiana
03 65 64 75 00 00 01 00 01 c0 0c 00 05 00 01 00   .edu...........
00 c3 42 00 17 05 73 74 6f 6e 65 03 75 63 73 07   ..B...stone.ucs.
69 6e 64 69 61 6e 61 03 65 64 75 00 c0 35 00 01   indiana.edu..5..
00 01 00 00 c3 42 00 04 81 4f 0c 01 c0 3b 00 02   .....B...O...;..
00 01 00 00 c3 b9 00 09 06 69 75 67 61 74 65 c0   ....iugate.
3b c0 3b 00 02 00 01 00 00 c3 b9 00 0b 05 6d 6f   ;.;...........mo
6f 73 65 02 63 73 c0 3f c0 3b 00 02 00 01 00 00   ose.cs.?.;......
c3 b9 00 11 05 61 72 67 75 73 03 63 73 6f 04 75   .....argus.cso.u
69 75 63 c0 47 c0 68 00 01 00 01 00 02 9b 0f 00   iuc.G.h.........
04 81 4f 01 09 c0 7d 00 01 00 01 00 02 9b 0f 00   ..O...}.........
04 81 4f fe bf c0 94 00 01 00 01 00 03 f4 20 00   ..O........... .
04 80 ae 05 3a   ....:
ID:768   0x8081   frame.relay.indiana.edu   CNAME   1 0xc3420017 stone.ucs.indiana.edu
ID:768   0x8081   frame.relay.indiana.edu   ANS   1 0xc3420004 129.79.12.1

```
TCP 1099 -> 80 seq:00000000 SYN  wnd:4096 options:04 02 B4 05 20 2F 54 48 size:4
IP 129.79.12.1 -> 204.148.87.155, Length 40, Protocol 6
TCP 1099 <- 80 seq:5ECF7C00 SYN ACK  wnd:4096 ack:00000001 size:0
TCP 1099 -> 80 seq:00000001 ACK PSH  wnd:4096 ack:5ECF7C01 size:493
IP 129.79.12.1 -> 204.148.87.155, Length 40, Protocol 6
TCP 1099 <- 80 seq:5ECF7C01 ACK  wnd:3603 ack:000001EE size:0
TCP 1099 -> 80 seq:000001EE ACK  wnd:4096 ack:5ECF7C01 size:0
IP 129.79.12.1 -> 204.148.87.155, Length 216, Protocol 6
TCP 1099 <- 80 seq:5ECF7C01 ACK PSH  wnd:4096 ack:000001EE size:176
TCP 1099 -> 80 seq:000001EE ACK  wnd:4096 ack:5ECF7CB1 size:0
IP 129.79.12.1 -> 204.148.87.155, Length 552, Protocol 6
TCP 1099 <- 80 seq:5ECF7CB1 ACK  wnd:4096 ack:000001EE size:512
TCP 1099 -> 80 seq:000001EE ACK  wnd:3584 ack:5ECF7EB1 size:0
IP 129.79.12.1 -> 204.148.87.155, Length 552, Protocol 6
TCP 1099 <- 80 seq:5ECF7EB1 ACK  wnd:4096 ack:000001EE size:512
TCP 1099 -> 80 seq:000001EE ACK  wnd:4096 ack:5ECF80B1 size:0
IP 129.79.12.1 -> 204.148.87.155, Length 552, Protocol 6
TCP 1099 <- 80 seq:5ECF80B1 ACK  wnd:4096 ack:000001EE size:512
TCP 1099 -> 80 seq:000001EE ACK  wnd:3584 ack:5ECF82B1 size:0
IP 129.79.12.1 -> 204.148.87.155, Length 552, Protocol 6
TCP 1099 <- 80 seq:5ECF82B1 ACK  wnd:4096 ack:000001EE size:512
IP 129.79.12.1 -> 204.148.87.155, Length 552, Protocol 6
TCP 1099 <- 80 seq:5ECF84B1 ACK  wnd:4096 ack:000001EE size:512
IP 129.79.12.1 -> 204.148.87.155, Length 552, Protocol 6
TCP 1099 <- 80 seq:5ECF7EB1 ACK  wnd:4096 ack:000001EE size:512
TCP 1099 -> 80 seq:000001EE ACK  wnd:3072 ack:5ECF86B1 size:0
IP 129.79.12.1 -> 204.148.87.155, Length 552, Protocol 6
TCP 1099 <- 80 seq:5ECF82B1 ACK  wnd:4096 ack:000001EE size:512
IP 129.79.12.1 -> 204.148.87.155, Length 552, Protocol 6
TCP 1099 <- 80 seq:5ECF84B1 ACK  wnd:4096 ack:000001EE size:512
IP 129.79.12.1 -> 204.148.87.155, Length 552, Protocol 6
TCP 1099 <- 80 seq:5ECF86B1 ACK  wnd:4096 ack:000001EE size:512
IP 129.79.12.1 -> 204.148.87.155, Length 552, Protocol 6
TCP 1099 <- 80 seq:5ECF88B1 ACK  wnd:4096 ack:000001EE size:512
TCP 1099 -> 80 seq:000001EE ACK  wnd:3072 ack:5ECF8AB1 size:0
IP 129.79.12.1 -> 204.148.87.155, Length 552, Protocol 6
TCP 1099 <- 80 seq:5ECF8AB1 ACK  wnd:4096 ack:000001EE size:512
IP 129.79.12.1 -> 204.148.87.155, Length 552, Protocol 6
TCP 1099 <- 80 seq:5ECF8CB1 ACK  wnd:4096 ack:000001EE size:512
IP 129.79.12.1 -> 204.148.87.155, Length 552, Protocol 6
TCP 1099 <- 80 seq:5ECF8EB1 ACK  wnd:4096 ack:000001EE size:512
TCP 1099 -> 80 seq:000001EE ACK  wnd:2560 ack:5ECF90B1 size:0
IP 129.79.12.1 -> 204.148.87.155, Length 552, Protocol 6
TCP 1099 <- 80 seq:5ECF90B1 ACK  wnd:4096 ack:000001EE size:512
TCP 1099 -> 80 seq:000001EE ACK  wnd:3584 ack:5ECF92B1 size:0
IP 129.79.12.1 -> 204.148.87.155, Length 552, Protocol 6
TCP 1099 <- 80 seq:5ECF92B1 ACK  wnd:4096 ack:000001EE size:512
IP 129.79.12.1 -> 204.148.87.155, Length 552, Protocol 6
TCP 1099 <- 80 seq:5ECF94B1 ACK  wnd:4096 ack:000001EE size:512
TCP 1099 -> 80 seq:000001EE ACK  wnd:3072 ack:5ECF96B1 size:0
IP 129.79.12.1 -> 204.148.87.155, Length 552, Protocol 6
TCP 1099 <- 80 seq:5ECF90B1 ACK  wnd:4096 ack:000001EE size:512
TCP 1099 -> 80 seq:000001EE ACK  wnd:4096 ack:5ECF96B1 size:0
IP 129.79.12.1 -> 204.148.87.155, Length 552, Protocol 6
```

TCP 1099 <- 80 seq:5ECF92B1 ACK wnd:4096 ack:000001EE size:512
IP 129.79.12.1 -> 204.148.87.155, Length 552, Protocol 6
TCP 1099 <- 80 seq:5ECF94B1 ACK wnd:4096 ack:000001EE size:512
IP 129.79.12.1 -> 204.148.87.155, Length 552, Protocol 6
TCP 1099 <- 80 seq:5ECF96B1 ACK wnd:4096 ack:000001EE size:512
IP 129.79.12.1 -> 204.148.87.155, Length 169, Protocol 6
TCP 1099 <- 80 seq:5ECF98B1 FIN ACK PSH wnd:4096 ack:000001EE size:129
TCP 1099 -> 80 seq:000001EE ACK wnd:3455 ack:5ECF9933 size:0
TCP 1099 -> 80 seq:000001EE FIN ACK wnd:4096 ack:5ECF9933 size:0
IP 129.79.12.1 -> 204.148.87.155, Length 552, Protocol 6
TCP 1099 <- 80 seq:5ECF96B1 ACK wnd:4096 ack:000001EE size:512
TCP 1099 -> 80 seq:000001EE FIN ACK wnd:4096 ack:5ECF9933 size:0
IP 129.79.12.1 -> 204.148.87.155, Length 40, Protocol 6
TCP 1099 <- 80 seq:5ECF9933 ACK wnd:4095 ack:000001EF size:0

TRACE 3
M - ES REGISTRATION, PINGS, CHANNEL HOP, DE-REGISTRATION

Reverse channel traffic appears in left hand column. Reverse channel data (M-ES to MD-IS) is displayed starting from the higher layer and ending with the lower layer. Forward channel traffic is tabbed towards the center, and is displayed in order from the lower layer to the higher layer.

Analysis Turned On at   08:46:32.00 on 10-13-95

-----Switching to channel: 717 at 08:46:32.82 on 10-13-95
Forward Sync Found at   08:46:32.82 on 10-13-95

UI TEI[1]= 0 C P=0 LMEI= f MsgType=Id Req 52 69 72 2e 30 30   8:46:54.55
 Vers=2 TxWin=15 RxWin=15 T200=30 N200=3 T203=0 P0=0 P1=2048 P2=16
(M-ES sends TEI assignment request)

          UI TEI[1]= 0 C P=0 LMEI= f   8:46:54.66
          MsgType=Id Assign 52 69 72 2e 30 30 TEI[3]= 07 d0 00
          Vers=2 TxWin=15 RxWin=15 T200=30 N200=3
          T203=0 N204=4 T205=5 P0=0 P1=2048 P2=16
          (MD-IS responds with TEI assignment with link level and compression parameters)

SABME TEI[3]= 07 d0 00 C P   8:46:54.66
(M-ES attempts to initialize MDLP)

          UA TEI[3]= 07 d0 00 R F=1   8:46:54.77
          (MD-IS acknowledges MDLP initialization)

          IFRAME NS= 0 NR= 0 TEI[3]= 07 d0 00 C P=0 Len=105   8:46:54.82
          SME M=0 K=1 CT= IP   8:46:54.82
           IKE K=0 ENC=TEST KeyBits=0
          **(MD-IS initiates key transfer. Note this is in "test" mode, and no real key exchange is
occurring.)**

SME M=0 K=1 CT= IP   8:46:54.82
  EKE K=0
(M-ES completes key exchange.)
IFRAME NS= 0 NR= 1 TEI[3]= 07 d0 00 C P=0 Len=40   8:46:54.82
(Info frame carrying SME from M-ES, above.)

MNRP M=0 K=0 CT= IP   8:46:54.82
  ESH AFI=49 IDI=0 DFI=0 IP=192.100.100.200 RegCount=1
  EquipId=52 69 72 2e 30 30
  Authentication Parameter:
   00 00 00 00 00 00 00 00 00 00
(M-ES sends End System Hello during registration.)
IFRAME NS= 1 NR= 1 TEI[3]= 07 d0 00 C P=0 Len=37   8:46:54.82
(Info frame carrying ESH from M-ES, above.)

        IFRAME NS= 1 NR= 2 TEI[3]= 07 d0 00 C P=0 Len=30   8:46:55.04
         MNRP M=0 K=1 CT= IP   8:46:55.04
          ISC AFI=49 IDI=0 DFI=0 IP=192.100.100.200
           ConfigTimer=28000
           Authentication Update Parameter:
            00 40 35 33 23 78 27 31 0e
        (Info frame from MD-IS carrying registration confirmation message.)

RR NR= 2 TEI[3]= 07 d0 00 R F= 0   8:46:55.54
(Link level ack sent by M-ES to previous forward channel NS=1 nfo frame.)

        IFRAME NS= 2 NR= 2 TEI[3]= 07 d0 00 C P=0 Len=134   8:47:03.57
        PROT=ICMP S=192.100.100.100 D=192.100.100.101 ID=240 Len=378
        IP M=1 K=1 CT= IP Len=129   8:47:03.57
        IFRAME NS= 3 NR= 2 TEI[3]= 07 d0 00 C P=0 Len=134   8:47:03.62
         IP M=1 K=1 CT= IP Len=129   8:47:03.62
        IFRAME NS= 4 NR= 2 TEI[3]= 07 d0 00 C P=0 Len=128   8:47:03.68
         IP M=0 K=1 CT= IP Len=123   ⸴ 8:47:03.68

        (**Three information frames sent from MD-IS. The first two indicate "more" segments to**
follow, while the last frame indicates "no more" segments to follow. This is an example of SNDCP
segmenting an ICMP "ping" message sent from the network towards the mobile user.)

RR NR= 5 TEI[3]= 07 d0 00 R F= 0   8:47:04.06
(MDLP response to NS=2, NS=3, NS=4, above.)

IFRAME NS= 2 NR= 5 TEI[3]= 07 d0 00 C P=0 Len=134   8:47:04.12
IP M=1 K=0 CT= IP Len=129   8:47:04.12
  PROT=ICMP S=192.100.100.101 D=192.100.100.100 ID=235 Len=378
IFRAME NS= 3 NR= 5 TEI[3]= 07 d0 00 C P=0 Len=134   8:47:04.12
 IP M=1 K=0 CT= IP Len=129   8:47:04.12
IFRAME NS= 4 NR= 5 TEI[3]= 07 d0 00 C P=0 Len=128   8:47:04.12
IP M=0 K=0 CT= IP Len=123   8:47:04.12
 (Three info frames on reverse channel carrying segmented echo response messages as a result of
receiving the "ping" sent from network to mobile.)

        RR NR= 5 TEI[3]= 07 d0 00 R F= 0   8:47:04.78
        (MD-IS with link level "ak" of received information NS=3, NS=3, NS=4.)

IFRAME NS= 5 NR= 5 TEI[3]= 07 d0 00 C P=0 Len=134   8:47:15.23
IP M=1 K=1 CT= IP Len=129   8:47:15.23
IFRAME NS= 7 NR= 5 TEI[3]= 07 d0 00 C P=0 Len=128   8:47:15.28
IP M=0 K=1 CT= IP Len=123   8:47:15.50
   PROT=ICMP S=192.100.100.100 D=192.100.100.101 ID=241 Len=378
SREJ NR= 6 TEI[3]= 07 d0 00 R F= 0   8:47:15.28

(An example of a selective reject. Notice information NS=5 was followed by NS=7. Accordingly, M-ES requests resequencing via selective reject NR=6.)

IFRAME NS= 6 NR= 5 TEI[3]= 07 d0 00 C P=0 Len=134   8:47:15.50
IP M=1 K=1 CT= IP Len=129   8:47:15.50
(Retransmitted frame, NS=6. This is another series of segments carrying an ICMP "ping".)

IFRAME NS= 5 NR= 8 TEI[3]= 07 d0 00 C P=0 Len=134   8:47:15.94
IP M=1 K=0 CT= IP Len=129   8:47:15.94
 PROT=ICMP S=192.100.100.101 D=192.100.100.100 ID=236 Len=378
IFRAME NS= 6 NR= 8 TEI[3]= 07 d0 00 C P=0 Len=134   8:47:15.94
IP M=1 K=0 CT= IP Len=129   8:47:15.94
IFRAME NS= 7 NR= 8 TEI[3]= 07 d0 00 C P=0 Len=128   8:47:15.94
IP M=0 K=0 CT= IP Len=123   8:47:15.94

(Three segmented echo responses to the received "ping," above.)

RR NR= 8 TEI[3]= 07 d0 00 R F= 0   8:47:16.60
(Link level "ak" for forward channel NS=5, NS=6, Ns=7.)

RRM Frame:107   8:47:30.46
 LMEI:42 Type=ChanStreamID CSI=0 Vers=1 Avail Hopping
 CellId=47 54 0 1a SPI=0 19 WASI=0 32
 PowerProd=35 MaxPower=0
(Forward channel issuing a Channel Identification message. Notice for Radio Resource management messages, this display does not show link level activity.)

RRM Frame:108   8:47:30.46
 LMEI:42 Type=CellConfig CellId=47 54 0 1a
 AreaColor=1 ActiveCSIs=1 RefChan=717 Face=0
 ERP Delta=0 RSSI Bias=0 PowerProd=35 MaxPower=0
 Channels:
  717
(RF cell configuration list for cell 1a.)

RRM Frame:109   8:47:30.46
 LMEI:42 Type=CellConfig CellId=47 54 0 16
 AreaColor=1 ActiveCSIs=2 RefChan=727 Face=0
 ERP Delta=0 RSSI Bias=0 PowerProd=35 MaxPower=1
 Channels:
  727 731
(RF cell configuration list for cell 16.)

RRM Frame:110   8:47:30.46
 LMEI:42 Type=ChanQuality RSSI_HYSTER=4 RSSI_SCAN_TIME=90
 RSSI_SCAN_DELTA=8 RSSI_AVG_TIME=5
 BLER_THRESH=20 BLER_AVG_TIME=5

(RF channel quality parameter message for the channel currently acquired.)

RRM Frame:116   8:47:34.37
  LMEI:42 Type=ChanAccess MAX_TX_ATTEMPTS=13 MIN_IDLE_TIME=2
    MAX_BLOCKS=64 MAX_ENT_DELAY=35 MIN_COUNT=5 MAX_COUNT=8
**(RF channel access parameters describing DSMA/CD MAC layer attributes. Notice that the
frame numbers went from 110 to 116. This is due to the operator halting the scope's run mode
temporarily, and shouldn't be considered a link level problem.)**

MNRP M=0 K=0 CT= IP   8:48:00.05
  ESB AFI=49 IDI=0 DFI=0 IP=192.100.100.200
    EquipId=52 69 72 2e 30 30
IFRAME NS= 8 NR= 8 TEI[3]= 07 d0 00 C P=0 Len=23   8:48:00.05
(Gracefull de-registration of M-ES.)

        RR NR= 9 TEI[3]= 07 d0 00 R F= 0   8:48:00.66
        (Link level "ak" for NS=8.)

DISC TEI[3]= 07 d0 00 C P=1   8:48:03.02
        DISC TEI[3]= 07 d0 00 C P=1   8:48:03.13
UA TEI[3]= 07 d0 00 R F=1   8:48:03.13
        UA TEI[3]= 07 d0 00 R F=1   8:48:03.13
(Both M-ES and MD-IS issues link level disconnects to each other, and both respond to each other.)

Analysis Turned Off at 08:48:08.96 on 10-13-95

Glossary

A carrier Typically, two cellular providers serve each market. The A carrier is associated with the cellular enterprise belonging to the local exchange carrier in that market, while the B side is typically the competing carrier. Each carrier is assigned its own frequency space within which to operate.

A interface The radiofrequency segment between a mobile user's CDPD modem (M-ES) and a CDPD service provider's cell site controller (MDBS).

Access control The mechanisms that define when and how a user may access a medium type without affecting other users. CDPD uses DSMA/CD as the access control mechanism on RF reverse channels. May also apply verification of passwords and other credentials required to use network application or network services.

Accounting The collection of variables used for billing and management purposes.

Accounting collector Consolidates accounting information received from distributors.

Accounting distributor Receives accounting information from accounting meters which are co-located at MD-IS sites. May also collect accounting information from other distributors. This information is then forwarded to other accounting collectors. Enables all networks and areas concerned with user traffic to become aware of use for billing and revenue-sharing purposes.

Accounting meter Measures user data transmissions for accounting purposes. Resident at MD-IS locations.

Accounting traffic matrix The collection of information that is measured by accounting meters. Contains statistics such as packets sent in either direction, registrations, etc.

Acquisition The process of an M-ES searching for, finding, and synchronizing to a DSMA/CD channel.

ACSP (AMPS channel status protocol) A CDPD-defined protocol that facilitates the exchange of information between the AMPS and CDPD environments. This information enables each system to become aware of resources that are in use, and pending use. This knowledge makes it possible to share hopping channels in a more graceful manner than RF sniffing.

Activity A term used to describe a peer-to-peer dialog at the application level.

Activity management   The responsibility of the session level in the OSI model. Activity management consists of sync pointing information transfers and the policing of transmissions via the exchange of tokens.

Adjacent cell   Cells located in such a way that an M-ES may learn of their existence via configuration messages and maintain service by hopping to them.

Adjacent MD-IS   MD-ISs are considered to be adjacent if at least one cell in each set of cells controlled by an MD-IS is adjacent to a cell which is under the control of a second MD-IS.

Adjacent MDBS   The radio controllers associated with adjacent cells.

Adjacent neighbor   Cells that are close enough to each other to allow an M-ES to maintain continuity of service when an unassisted channel hop occurs.

Administrative domain   The area(s) under control of a specific CDPD carrier, in which the carrier is responsible for administration of management, billing, and network services.

Advanced Mobile Phone Service   The North American analog standard for mobile telephone use.

AFI (authority and format identifier)   Part of an OSI address which identifies the authority and format of a variable-syntax NSAP address.

Air link interface   The segment between the CDPD service provider and an M-ES. May also include cell site-to-MD-IS facilities.

Allocated channel   A channel that is CDPD capable.

AMPS   See **Advanced Mobile Phone Service.**

Application entity   A process being accessed that resides in end systems and provides a service to the user. E-mail, network management, and billing processes are examples of application entities.

Application services   The services that are provided by an application entity. End-system resident.

Area color code   A 3-bit field within a Reed Solomon block that identifies an MD-IS in control of a channel stream.

ARN   See **Authentication random number.**

ARQ (automatic retransmission request)   The act of asking for a retransmission of a frame which has been deemed to be corrupt due to bit errors.

ASCII (American Standard Code for Information Interchange)   A 7-bit character set, also known as IA5.

ASN   See **Authentication sequence number.**

ASN.1   An international protocol which allows for the nonambiguous identification of data formats and syntax.

AT   The "attention" command to enter an administrative dialog with modems and other communications devices that are Hayes compatible.

ATM   A networking technology which utilizes cell switching technology. Cells are fixed-length (53-byte) data units. Because they are fixed in length, hardware switching solutions can apply, increasing switching speeds dramatically.

Authentication   The process of validating a CDPD mobile user as a bona-fide user of network services.

Authentication random number   A random number generated by home MD-IS entities which is maintained by M-ES equipment. The number is regenerated with each successful registration.

Authentication sequence number   A number which is initially generated by random home MD-IS entities and is maintained by M-ES equipment. The number is incremented by a value of 1 with each successful registration.

Authority and format identifier   See **AFI.**

Available channel   A channel that can accept additional registrations for packet use.

AVL   Automatic vehicle location.

B carrier   Typically, two cellular providers serve each market. The A carrier is associated with the cellular enterprise belonging to the local exchange carrier in that market, while the B side is typically the competing carrier. Each carrier is assigned its own frequency space within which to operate.

Backbone   A term commonly used to describe communications facilities that do not connect directly to end-user systems. A backbone typically interconnects hubs, which in turn connect to the end-user population.

BCD (binary-coded decimal)   A means to express the values 0 through 15 by individual symbols ranging from 0 through F.

BER   See **Block error rate.**

BGP   See **Border gateway protocol.**

Bit stream   The sequence of binary information transmitted over a serial interface or transmission facility, such as CDPD forward and reverse channels.

Bit transmission rate   The number of binary bits that are transmitted in a 1-second interval within a bit stream.

BLER   See **Block error rate.**

Block   With respect to CDPD, a block is a 63–47 Reed Solomon data unit.

Block error rate   The ratio of blocks which cannot be corrected to blocks received that have no bit errors, or that can be corrected.

Border Gateway Protocol   An exterior gateway protocol defined in RFCs 1267 and 1268. Its design is based on experience gained with EGP (Exterior Gateway Protocol), as defined in STD18, and RFC904.

Broadcast   A transmission which is directed to all stations listening, as opposed to a point-to-point, directed transmission.

BSC (base station controller)   Resident within a cell site, the BSC manages RF channel resources.

Busy state   The state of a reverse channel stream as indicated in a forward channel stream's busy/idle status flag.

Busy/idle flag   A 5-bit sequence transmitted within the forward DSMA channel stream which indicates if a station is currently making use of the corresponding reverse channel within the channel pair.

CCITT (International Telegraph and Telephone Consultative Committee) United Nations treaty organization responsible for international standardization of various communications and network disciplines. Now referred to as ITU, International Telecommunications Union.

CDMA (code division multiple access)   A medium access mechanism which supports multiplexing capability by use of spread spectrum technology.

CDPD (cellular digital packet data)   Mobile IP over cellular radio systems.

CDPD cell boundary   The point at which received signal strength (RSSI) derived from a specific cell is inadequate to maintain continuity of service for an M-ES.

CDPD service provider   An organization that is responsible for providing, managing, supporting, and billing CDPD services for its customer base.

CDPD service provider network   The network operated by the CDPD service provider.

CDPD SNDCP   The CDPD subnetwork-dependent convergence protocol. Responsible for the encryption and compression of user data before transmission over an air interface.

Cell   The area covered which provides a signal on a specific frequency which is adequate for an M-ES to maintain continuity of service.

Cell group color code   A 5-bit sequence found within Reed Solomon blocks that identifies a set of cells, which are arranged within a cluster. The frequencies used within the cluster can then be reused at distant locations.

Cell transfer   The act of an M-ES switching to another channel stream which is being provided by another cell.

Cellular digital packet data   See **CDPD.**

Center frequency   The frequency associated with a given cellular channel.

CGSA (cellular geographic service area)   The set of cells under the administration of a network service provider.

Channel acquisition   See **Acquisition.**

Channel bandwidth   The information-carrying capacity of a channel. In analog terms, the difference between the highest and lowest frequencies on a channel; 30 kilohertz for CDPD channels. In digital terms, the number of bits per second that can pass over the channel.

Channel capacity   Taken in context, can refer to either the digital bandwidth of a channel or how many concurrent users may be supported (registered) on a channel.

Channel congestion   The point where acceptable levels of service cannot be met on a channel due to excessive contention between multiple users, or channel monopolization by an individual user.

Channel hop The act of an M-ES relinquishing use of a current channel and acquiring a new channel for use.

Channel sealing The act of taking a CDPD channel out of service.

Channel stream A Reed Solomon–encoded bit stream which carries CDPD traffic over an AMPS channel. A forward and a reverse channel stream are required for M-ES communications; referred to as a channel "pair."

Client Typically, a computer on the user side of a network connection which requests services from a remote server. The server provides information to the client; the client processes the information it receives.

CLNP (connectionless network protocol) An ISO 8473–defined protocol which provides a connectionless datagram delivery service on behalf of users.

CLNS (connectionless network service) The type of service provided by CLNP and IP protocols. Datagram in nature, there are no formal session establishment handshakes or acknowledgments.

CLS See **CLNS.**

Clone An impostor device which is using authentication credentials in an illegal manner.

CM-ES (circuit switched mobile-end system) An M-ES which has the ability to access the CDPD network via hybrid CS-CDPD access procedures.

CMD-IS (circuit switched mobile data intermediate system) An MD-IS which supports circuit switched access via hybrid CS-CDPD procedures.

CMIP (Common Management Information Protocol [CCITT X.711]) The protocol and PDU structure for OSI-based network management information exchanges.

CMIS (Common Management Information Service [CCITT X.710]) The OSI-defined network management service.

Co-channel interface Interference from a remote cell on frequencies being used in another cell.

Color code The 8-bit word found in Reed Solomon blocks which defines an MD-IS service area in 3 bits and a cluster of cells known as a cell group in 5 bits.

Compression The act of eliminating redundancies in data transmissions in an attempt to reduce the overall number of bits transmitted in an output burst. V.42bis defines one standard approach to data compression.

Confidentiality Keeping user information secure by way of encryption or other methods.

Configurable parameter A user or network variable that can be defined, usually associated with "fine-tuning" a network environment.

Confirmation A primitive that informs a requesting party that the request is being processed. Passed from an N layer to an $N + 1$ layer.

Control information Overhead associated with a protocol layer within a communications profile. Also known as "header" information.

Correctable error   A bit error that can be detected and corrected by the receiving device via parity and other algorithms. Having the ability to correct bit errors reduces the need for retransmission requests.

COS (connection-oriented service)   A network service that utilizes virtual circuits, housekeeping activities, or both. Also, Corporation for Open Systems.

COTS   A connection-oriented transport service such as is delivered by the DOD TCP and ISO TP4 protocols.

CRC (cyclic redundancy check)   An algorithm based on properties of prime numbers that assists in the identification of bit errors.

Credentials   The set of variables that are customer specific and uniquely identify a CDPD subscriber as being a bona-fide user of network services.

CS-CDPD (circuit switched cellular digital packet data)   A hybrid service that enables wireline or AMPS circuit switched access to the CDPD network.

CSCCP (circuit switched CDPD control protocol)   A CS-CDPD function that emulates the MDLP service for SNDCP and supports circuit mode connection procedures for hybrid service.

CSI (channel stream identifier)   The carrier-assigned number identifying a CDPD channel stream.

CTIA   Cellular Telephone Industry Association.

D channel   An out-of-band signaling channel used in the ISDN environment which supports message-oriented signaling via the LAPD and Q.931 protocols.

Datagram   A packet that is delivered via a connectionless network.

dB   Decibels. A logarithmic expression of power differences between two sources.

dBm   A logarithmic expression of power differences between two sources, with 0 milliwatts as the reference point.

dBW   A logarithmic expression of power differences between two sources, with 0 watts as the reference point.

DCC (data country code)   An internationally defined set of digits which uniquely identifies a national plan for address administration relating to NSAPs.

DCE (data communications equipment)   Modems, multiplexers, and the like, which connect terminal devices to transmission facilities. Also refers to network switching elements.

Decibel   A value expressed in decibels is 10 times the logarithm of power reference 1 over power reference 2.

Decompression   The act of extracting an original data stream by invoking a data compression algorithm in reverse. In the case of V.42bis, this is done by table lookups which substitute character strings for smaller "codewords."

Decryption   The act of extracting an original data stream that has been modified according to a specific algorithm.

Decryption key Defines procedures for a decryption algorithm to identify and extract a modified data stream.

Dedicated channel In CDPD, a channel that is not known by the AMPS system and is available only for CDPD use.

Demultiplexing The act of extracting an individual channel stream from a composite stream consisting of more than one channel of information. This can be done by frequency assignments, time slot assignments, or via logical address assignments, as is the case with CDPD.

Deregistration The act of an M-ES "gracefully" informing a CDPD network that services are no longer required. This causes MD-IS location and forwarding tables to be flushed.

DFI (domain format identifier) Used within the NSAP address. In CDPD has a value of 128 decimal, which correlates to US GOSIP–formatted NSAP address structure.

Digital-sense multiple access with collision detection (DSMA/CD) The CDPD-specific RF channel access mechanism, which is based on a contention-resolution scheme.

Directed hop An M-ES's move to a new CDPD channel in response to a directive received by the CDPD network.

Directory service A service that performs name-to-address translation upon user request.

DISC (disconnect) A standard HDLC (high-level data link control) verb for releasing a data link connection between two endpoints or network nodes.

DLC (data link control) A generic term for layer 2 hop-to-hop management.

DLCI (data link connection identifier) The name for a logical connection between two link level nodes. In CDPD, DLCIs are identified with the MDLP address, or TEI.

DM (disconnect mode) A standard HDLC verb for identifying that an interface on a network node is not active.

DOD U.S. Department of Defense.

Domain format identifier (DFI) Part of an NSAP address. Specifies the format of the DSP part of the address. CDPD uses a value of 128 (80 hex) to identify that the DSP will use GOSIP v.2 formats.

DS0 (digital signal level 0) A 64,000-bit-per-second digital channel consisting of 8 bits sent at 8 kilohertz.

DS1 (digital signal level 1) A 1,544,000-bit-per-second data stream which is framed according to ATT 62411. Accommodates 24 DS0s.

DS3 (digital signal level 3) A 44,736,000-bit-per-second data stream framed according to ATT 62415. Accommodates 28 DS1s.

DSMA/CD See **Digital-sense multiple access protocol with collision detection.**

DSP (domain-specific part)   Part of an NSAP address which is administered by a unique authority, such as ANSI.

DTE (data termination equipment)   The source and destination of user data. In CDPD, the DTEs are end systems.

Duty cycle   The ratio of time "in use" to time not "in use."

E interface (external interface)   An interface between a CDPD service provider's network and another (CDPD or private) network.

EIA   Electronic Industries Association.

EID (equipment identifier)   In CDPD, a 48-bit number assigned by the IEEE.

EIRP   Effective isotropic radiated power.

Electronic key exchange   An automated process which allows real-time rekeying of an encryption algorithm. In CDPD, this is performed upon registration of an M-ES.

Encapsulation   The act of taking a unit of data and placing that data within another packet.

Encryption   The act of making a data transmission secure by modifying the transmitted bit stream with a pseudo-random pattern as defined by a mathematical algorithm.

Encryption key   Defines procedures for an encryption algorithm to modify an original data stream.

End system (ES)   Data processing systems that provide application-level services for user communities.

Entrance delay   The length of time in microslots an M-ES must wait before attempting to access a reverse channel.

ERP   Effective radiated power.

ES   See **End system.**

ES-IS (end system–intermediate system)   An internationally defined routing protocol.

ESB (end system bye)   Used with MNRP to deregister an M-ES.

ESH (end system hello)   A message included in the mobile network registration protocol and sent when an M-ES wishes to register for service.

ESQ (end system query)   A message that is part of the MNRP message set and is issued by the serving MD-IS when attempting to verify the presence of an NEI.

External F-ES   A fixed-end system that performs application-level services, but is not under the administration of the CDPD carrier.

F-69   The international telex numbering plan.

F-ES   See **Fixed-end system.**

Face   Part of a cellular antenna assembly which radiates in a directional manner.

Face neighbor   Cells that are on adjacent faces.

FCS (frame check sequence)   A sequence of bits, typically 16 or 32 bits, which enables a device to determine if received data units contain errored bits.

FEC (forward error correction)   The act of detecting, and correcting, bits which have been modified during transmission at the receiver's end. This eliminates the need for retransmission requests.

Fixed-end system (F-ES)   An end system that provides application-level services, but is not mobile.

Flag   A bit sequence of 0 1 1 1 1 1 1 0 which serves as the starting and ending delimiters for bit-oriented protocols such as MDLP.

Flow control   The act of being able to force a reduction or increase of output data from a remote system's transmitter.

Forced hop   An M-ES's move to a different CDPD channel stream due to unacceptable operating characteristics on a current CDPD channel stream.

Forward channel   The "downlink" or channel stream in the MDBS-to-M-ES direction.

Forward error correction   See **FEC**.

Frame   The generic name for a unit of data created by a layer 2, or link level protocol such as MDLP.

Frame relay   A popular link level protocol which provides frame switching services in a virtual circuit environment.

FRMR (frame reject)   An HDLC verb which signifies unrecoverable link level problems.

FTAM (file transfer and management)   An OSI application offering more comprehensive services than FTP.

FTP (file transfer protocol)   A file transfer application native to the TCP/IP suite.

Full-duplex M-ES   An M-ES which is capable of transmitting data in the reverse or uplink direction while simultaneously observing the forward channel stream.

Gaussian minimum shift keying (GMSK)   The modulation scheme used by CDPD which distinguishes a CDPD signal from an AMPS carrier.

GMID (multicast group member identifier)   All members of a common multicast group are associated with a common GMID.

GSM (groupe speciale mobile)   Part of the signaling structure commonly found within European cellular environments.

Half-duplex M-ES   An M-ES which does not have the ability to process the forward CDPD channel stream as it is transmitting a burst over the reverse channel stream.

HDLC (high-level data link control)   An industry-standard means to encapsulate data units between two network nodes or link level hops.

Header The protocol control information that is part of the overhead generated by any given protocol process within a communications profile.

HLR (home location register) Found within the AMPS environment, this is analogous to the MHF of CDPD.

Hop channel The "target" channel of an M-ES when a move to another channel stream is required.

Housekeeping A term to describe the set of procedures for acknowledgment, flow control, and other connection establishment and management routines that are particular to connection-oriented services.

Hysteresis Continuous changing or toggling; an undesirable phenomena with respect to signal acquisition.

I interface (interservice provider interface) The IS-to-IS junction between two CDPD carriers.

IA5 (International Alphabet #5) See **ASCII.**

IDP (initial domain part) The combination of the AFI and IDI within NSAP addresses.

IDRP Inter-Domain Routing Protocol.

IEEE Institute of Electrical and Electronic Engineers.

IGOSS (Industry/Government Open System Specification) A collection of protocols and communications procedures that have been standardized for use by industry and government consortiums, with the ultimate goal of achieving standardization and systems interoperability.

Implementation dependent A term used for design criteria that are not standardized but are left to a vendor's creativity and interpretation.

In-band control The act of sending administrative or control information in the form of tones, codewords, or messages over the same channel space over which user applications communicate.

Indicate A primitive call passed from an N layer to an $N + 1$ layer to request services or inform an $N + 1$ layer of incoming data.

Interarea cell transfer Movement of an M-ES from a serving area controlled by one MD-IS to a serving area controlled by another MD-IS. Requires a new registration/ESH sequence from the mobile user.

Intermediate system (IS) A router which is responsible for the relaying of Internet datagrams.

Internal F-ES A system which provides application-level services and is under the administration of the CDPD carrier.

Internet The collection of transmission facilities, intermediate systems, and end systems that make up the network originally deployed by the Defense Advanced Research Project Agency (DARPA). Formally referred to as "the connected Internet."

Intraarea cell transfer A cell transfer which remains within an area served by an individual MD-IS.

IP   The term used for the DOD RFC791 Internet Protocol.

IS   Intermediate system, or router. Also, an acronym for International Standard.

IS-41 (Interim Standard 41)   A series of standards defined by EIA/TIA which address issues related to intersystem operations of a cellular network, such as automatic roaming capability and management.

IS-54 (Interim Standard 54)   A series of standards defined by EIA/TIA which address issues relating to dual-mode (AMPS/TDMA) systems compatibility.

IS-IS   Intermediate systems–to–intermediate systems routing protocol. An OSI standard.

ISC   MD-IS Hello Confirm packet. The expected response to an M-ES's end-system hello message, which is sent when attempting to register for network services.

ISDN   Integrated services digital network.

ISO   An abbreviation used on documents created by the International Organization for Standardization.

ISO DCC (ISO Data and Country Code)   A format for NSAP addresses as defined by ISO 8348.

ISP   Can stand for an International Standardized Profile such as the US GOSIP profile, or Internet service provider.

ITU   International Telecommunications Union. ITU-R follows radio issues, while ITU-T follows telecommunications issues.

K bit   Kilobit, or 1024-bit segment. Also, a bit that is toggled by the SME when electronic key exchanges occur.

K parameter   Layer 2 (HDLC) window size.

Kbps   Kilobits per second.

Key   See **Encryption key.**

Key exchange   The act of informing a device of a new encryption key to be used for security purposes.

Key generation   The rules for changing a key for encryption purposes.

Key management   The set of rules for defining when and how a new encryption key will be generated and exchanged.

KHz   Kilohertz, measured in thousands of cycles per second.

LAN   Local area network.

LAPB (Link Access Procedure B)   A balanced, connection-oriented version of HDLC.

LAPD (Link Access Protocol version D)   The layer 2 framing protocol used with ISDN signaling. CDPD's MDLP is a derivative of this.

LAPM   Link access procedure for modems. Also known as V.42.

Layer management entity identifier (LMEI)   A code used to determine what process is using the MDLP service.

LCI (local cell identifier)   A carrier-assigned number identifying a cell within its serving area.

LD (location directory)   A directory informing a mobile home function of where one of its client M-ESs is located when registered with the network.

Link layer   The layer 2 hop-to-hop control or framing level between two network nodes.

LMEI   See **Layer management entity identifier.**

Local cell identifier   See **LCI.**

Local service area identifier   A number which identifies a set of cells associated with common accounting procedures; part of R1.0, but has been removed from R1.1 of the CDPD standard.

Location directory   See **LD.**

LSAI   See **Local service area identifier.**

M bit   More segments. This bit indicates that more segments associated with a common PDU will follow.

M-ES (mobile-end system)   The CDPD mobile device which must register and be granted permission to use the network service after authentication has been verified.

MAC (medium access control)   A generic term for the criteria used to decide whether a transmission facility is busy or idle, and if it can be accessed without affecting other users in an adverse manner.

MAC layer   The process responsible for medium access and control decision making.

MAC protocol   A protocol implemented by the MAC layer. In CDPD this is called DSMA/CD, or digital-sense multiple access with collision detection.

Management   The activities associated with defining network configuration, maintenance, security, and performance.

MAS (mobile application subsystem)   Part of an M-ES which provides application-level services.

MD   Mobile data, as in mobile data base station (MDBS), mobile data intermediate system (MD-IS), etc.

MD-IS (mobile data intermediate system)   Provides serving functions to mobile-end systems and authentication or forwarding services for client M-ESs. Supports CDPD-specific mobility management services as well as basic routing (IS) and datagram-relaying services.

MD-IS serving area   The set of cells controlled by a single serving MD-IS.

MDBS (mobile data base station)   The CDPD channel stream controller.

MDLP (mobile data link protocol)   The HDLC LAPD derivative which is used between serving MD-IS and M-ES hops or data link connections.

Medium access control   See **MAC.**

Message   Any block of data created by a layer within a system profile. Also called a PDU, or protocol data unit.

Message handling system (MHS)   A term used to identify a system capable of store and forward delivery of messages. Typically associated with electronic mail systems.

MHF (mobile home function)   A function within every MD-IS which is responsible for authentication of its "client" M-ES population and forwarding of datagrams to M-ESs when registered.

MHS   See **Message handling system.**

MHz   Megahertz, measured in millions of cycles per second.

Microslot   The time between two consecutive busy/idle status flags within a Reed Solomon block. At 19,200 bits per second, this is equal to 3.125 milliseconds.

MIN   Mobile identification number.

MNLP (mobile network location protocol)   The protocol used between home and serving MD-IS functions during authentication and registration events.

MNRP (mobile network registration protocol)   The protocol used between serving MD-ISs and mobile-end systems during registration and deregistration events.

Mobile   An adjective describing a device as being nonstationary. A noun correlating to an end system which is nonstationary.

Mobile application subsystem (MAS)   See **MAS.**

Mobile data base station (MDBS)   See **MDBS.**

Mobile data intermediate system (MD-IS)   See **MD-IS.**

Mobile data link protocol (MDLP)   See **MDLP.**

Mobile-end system (M-ES)   See **M-ES.**

Mobile network location protocol (MNLP)   See **MNLP.**

Mobile network registration protocol (MNRP)   See **MNRP.**

Mobile serving function (MSF)   The function within an MD-IS which maintains a data link connection with a mobile device and acts as an intermediary between the mobile device and its home MD-IS during registration and authentication events.

Mobility management   The set of procedures required to authenticate, locate, and maintain continuity of service for mobile-end systems.

Modem (modulator/demodulator)   Modems take digital information and encode this information onto an analog carrier (frequency) by modulating the amplitude, phase, frequency or a combination of these attributes.

Modulation   The act of changing the phase, frequency, amplitude, or a combination of these attributes to encode digital information onto the original waveform, or "carrier."

MSA (metropolitan statistical [service] area) A geographic region which is serviced by a cellular carrier.

MSC (mobile switching center) The telephone office where cellular calls are managed. CDPD MD-IS may be co-located with AMPS switching systems in the MSC. Also referred to as MTSO, mobile telephone switching office.

MSF See **Mobile serving function.**

Multicast A form of broadcasting where only a select group of stations receives the message. Stations must be assigned as members of a common multicast group. For CDPD, this is done manually by coordinating service requirements with the service provider.

Multiplexing The act of combining more than one data stream onto a single transmission facility. This can be done using frequency assignments, time slot assignments, or logical addresses associated with each user. The latter, called statistical multiplexing, is the technology used by CDPD.

N200 An M-ES's maximum retransmission counter at the MDLP level.

N204 The maximum number of TEI notifications allowed at the link level.

NDIS (network driver interface specification) A standard software process that enables data transfer between a CPU and connected peripheral devices. Associated with network interface cards (NICs), which connect workstations to local area network (LAN) segments.

NEI (network entity identifier) IP or NSAP address associated with M-ESs, F-ESs, MD-ISs, and other network-addressable entities in the CDPD environment.

Network application services CDPD network-resident application processes, such as e-mail, billing, and directory services.

Network entity identifier (NEI) See **NEI.**

Network layer protocol identifier (NLPID) A code used to identify what network level protocol is being serviced. Typically found at the link level, but with CDPD, this code is found within the SNDCP header.

Network management system (NMS) A generic term for a system which enables network management functions (see **Network management**), typically from a remote location.

Network protocol data units (NPDUs) Messages generated by the network-level entity.

Network service access point (NSAP) A single-byte code found at the end of an NSAP address which identifies the higher-layer entity using the CLNP service. Also referred to as the OSI 8473 CLNP address.

NIC Network interface controller or network information center. An organization responsible for the administration of IP domain names and address assignments. For CDPD, Network Solutions, Inc., of Herndon, VA.

NIST National Institute for Standards and Technology (U.S.).

NLPID See **Network layer protocol identifier.**

Node An intermediary point within a network that is responsible for relaying information among individual network segments or hops.

Nondedicated channel   A channel that is allocated for both AMPS and CDPD traffic. Also referred to as a hopping channel.

NSAP address   See **Network service access point.**

NSAP selector   The single-byte code found at the end of an NSAP address.

OAM (operations and management)   An acronym used to describe processes that facilitate network management.

Object identifier (OID)   A string of integers which uniquely identifies a manageable device or process.

Octet   Eight consecutive bits.

ODI   One of many standard software processes that enable data transfer between a CPU and connected peripheral devices. Associated with network interface cards (NICs), which connect workstations to local area network (LAN) segments.

Off the shelf   A commonly used term to describe products that are commercially available.

Open data link interface   See **ODI.**

Open shortest path first   See **OSPF.**

OSI   Open Systems Interconnect.

OSPF   Open shortest path first. A routing protocol.

Packet   A message created by the network level within a communications system. Also called an NPDU.

Parent   In V.42bis, a node (character) that is directly above another in a dictionary's tree entry.

Parity symbol   A 6-bit symbol within a Reed Solomon block which is used for FEC purposes.

PBX (private branch exchange)   A privately owned voice or data telephone system.

PCI (protocol control information)   The overhead created by a given software process, also called "header" information.

PCS (personnal communication services)   A set of services which provides mobile voice and data capability within metropolitan areas using digital CDMA technology.

PDU (protocol data unit)   The message created by a given software process, including its own header.

Peer-to-peer communications   A term used to describe the cooperation of corresponding layers during communications between two systems.

Physical layer   The attributes of a machine's input and output functions. Physical layer definitions include the electrical, mechanical, and procedural elements of operation.

PIN   Personal identification number, used for security purposes.

Planned hop   A channel hop that is prearranged by time of day and defined by the CDPD services provider.

Point-to-multipoint   A topology which includes broadcast in one direction to a population of "remote sites" but point-to-point information flow in the reverse direction, back to the "central site." This is the topology of a CDPD RF segment, where the M-ESs are "remote sites" and an MD-IS is the "central site."

Point-to-point   Information flow between two, and only two, stations.

Port   In hardware, a physical serial interface. In software, the interface commonly associated between layers 4 and 5 within an application stack.

POS (point of sales)   Automated debit and credit applications, such as credit card or bank card reading.

Power adjustment   The act of an MDBS directing an M-ES to adjust (attenuate) its output power levels to an acceptable level. Managed by the CDPD radio resource management entity.

Power class   Standard output power levels for equipment categories according to EIA/TIA 553.

Power level   A specific output power level as defined by equipment class in EIA/TIA 553.

Power product   Part of an MDBS channel broadcast message that defines transmit-to-received power differences and is used by an M-ES to determine best channel selections.

PPP (point-to-point protocol)   A standard protocol which defines encapsulation methods for IP traffic as well as a PAP (password authentication protocol) and CHAP (challenge authentication protocol) set of procedures which are typically employed over switched access facilities.

Primitive   System calls between layers to request and deliver services across layer boundaries.

Profile   The set of protocols implemented in layers within a communications system.

Protocol   The elements of procedures, including message structure definitions, that are used between communicating systems.

Protocol control information   See **PCI.**

PSTN   Public switched telephone network.

PVC   Permanent virtual circuit.

QOS (quality of service)   Attributes of a communications dialog that pertain to reliability, throughput, transit delay, and other variables that affect users' perception of overall system performance.

R1.0   The first release of CDPD standards, dated July 19, 1983.

R1.1   Revision 2 of the CDPD standards, released January 19, 1995.

Radio resource management   The CDPD set of procedures that is responsible for channel acquisition, channel hopping, power-level output, etc.

Ramp-down   The time between the end of an M-ES's transmission burst and the reduction of power on the reverse channel so as to be undetectable by an MDB.

Ramp-up   The time for an M-ES's output power to reach acceptable signal strength before a reverse channel transmission burst begins.

RC4   An encryption algorithm defined by RSA Securities, Inc.

RDC (redirect confirm)   Part of the MNLP message set.

RDQ (redirect query)   Part of the MNLP message set.

RDR (redirect request)   Part of the MNLP message set.

Reassembly   The act of combining individual segments within a data stream so as to reproduce an original message which was larger than the capabilities of the transport services delivering the message.

Received signal strength indication (RSSI)   The power level of a received signal.

Reed Solomon code   An extremely powerful and reliable class of nonbinary BCH codes used for error correction purposes. A description of BCH codes can be found in S. Lin and D. J. Costello, Jr., *Error Control Coding: Fundamentals and Applications* (Prentice-Hall, Englewood Cliffs, NJ, 1983).

Registration   The act of a mobile device requesting service and authentication from a CDPD service provider.

REJ (reject)   A standard HDLC verb used for resequencing out-of-sequence information frames.

Reliable sequenced delivery   The form of delivery provided by CDPD's MDLP protocol. Information is sequenced and delivered in order across the air interface.

Reply   A service primitive sent from an $N + 1$ layer to an N layer after receipt of an indicate.

Request   A service primitive sent from an $N + 1$ layer to an N layer to initiate a service request.

Reverse channel   The channel stream used by an M-ES for transmission purposes.

RF   Radiofrequency.

RF channel   A transmission medium using radiofrequencies; for CDPD, the channel sets defined in EIA/TIA 553.

RF channel number   For CDPD, the 30-kilohertz-wide channels defined in EIA/TIA 553.

RF channel pair   The set of forward and reverse channels that facilitate communications to and from an M-ES.

RFC (request for comments)   Internet policies and guidelines, commonly considered to be Internet standards.

RIP   Routing Internet protocol.

RNR (receiver not ready) A flow control verb used within the HDLC message set, including CDPD's MDLP. This verb is used to stop transmission of information frames from the remote endpoint.

Roaming M-ES An M-ES being served outside its home service area.

Root The top node in a V.42bis dictionary entry.

Routing domain The set of intermediate systems under the authority of one CDPD carrier.

Routing protocol A protocol which enables intermediate systems to relay network topology information among each other.

RR (receiver ready) A flow control verb used within the HDLC message set, including CDPD's MDLP. Used to inform the remote endpoint that it may commence information transfer.

RRM See **Radio resource management.**

RRMP Radio resource management protocol.

RS-232 A commonly used serial interface specified by the Electronics Industry Association.

RSA (rural service area) A geographic region serviced by one or more cooperating cellular carriers.

RSSI See **Received signal strength indication.**

SABME (set asynchronous balanced mode extended) The link level initialization command used by HDLC if sequence numbering of information in the range 0–127 is desired. Applicable for CDPD's MDLP.

SABRE (Semi-Automatic Business Research Environment) A reservation and inventory control system used by American and other airlines.

SAP (service access point) A code that is assigned to higher-layer protocols and is used for directing messages to the appropriate software processes within a communications system.

SDU (service data unit) Data that has yet to be processed by a given layer entity. For example, when layer 4 hands information to layer 3, layer 3 considers this data an SDU.

Sector The area covered by an RF channel when radiated via a directional antenna.

Security The protection of data or other resources from eavesdropping, unauthorized access, or theft.

Segmentation The process of taking an original (large) message and fragmenting the message into smaller pieces before transmission over a communications channel. The individual segments are collected and reassembled at the distant endpoint.

Server Provides services to a client software process.

Service access point See **SAP.**

Service provider The organization responsible for operation of a CDPD network.

Service provider network identifier An ANSI-assigned number associated with a given CDPD service provider.

Serving MD-IS The MD-IS which provides the mobile serving function (MSF) for an M-ES.

SIM (subscriber identity module) Part of an M-ES which contains user-specific information associated with security and validation.

Sleep mode A mode of operation in which an M-ES's transmitter functions are suspended in an effort to conserve power sources.

SLIP (serial line IP) A protocol defined in RFC1055 which recommends a way to frame IP datagrams over a serial communications line.

SMDS (switched multimegabit data service) A connectionless high-speed delivery service defined by Bellcore.

SME The security management entity within the CDPD network that ensures the confidentiality of user data via encryption services.

SMP (security management protocol) The message set associated with CDPD's encryption services.

SMR (specialized mobile radio) Wireless communications systems often used for fleet management purposes and inventory control applications.

SMTP (simple mail transfer protocol) Part of the TCP/IP protocol suite.

SNDCF (subnetwork-dependent convergence function) A process which mates network and transport functions within a communications profile to link level and physical-layer processes. The function hides the details of lower-layer attributes from the network and transport layers. In CDPD, this function is provided by the SNDCP protocol.

SNDCP (subnetwork-dependent convergence protocol) A CDPD-specific process that provides encryption and compression on user data prior to transmission. SNDCP also performs segmentation and reassembly functions on datagrams being relayed over the air interface.

Sniffer A term used to describe an MDBS function which detects non-CDPD signals on an RF channel that is provisioned to support CDPD services. When non-CDPD signals are detected, MDBS transmission ceases so as not to create interference.

SNMP (simple network management protocol) An industry-standard protocol used for remote management of network elements and processes.

Socket A software interface between layers. Analogous to a SAP, or PORT.

SONET (synchronous optical network) A high-speed physical transport medium.

SPI Service provider identifier.

SPNI See **Service provider network identifier.**

SREJ (selective reject) An optional verb within the HDLC message set that enables the resequencing of individual frames which may have been deemed to be out of sequence.

Stack   The set of layers within a system profile.

SU (subscriber unit)   The part of an M-ES which supports CDPD-specific functions such as SNDCP, MNRP, etc.

Subarea routing domain   The set of cells under the control of an individual MD-IS.

Subnetwork-dependent convergence protocol   See **SNDCP.**

Subscriber application   The e-mail, inventory control, point-of-sale, etc., functions that are being put to work by an end user.

Subscriber identity module   See **SIM.**

Subprofile   A set of software services which define the transport and delivery attributes of communications services. A subprofile may consist of connectionless or connection-oriented transport and packet delivery mechanisms.

SVC   Switched virtual circuit.

Symbol   A 6-bit word created by the Reed Solomon blocking mechanism.

Sync pointing   A form of activity management performed by the session level. Resynchronizes connections after a network failure.

Synchronization   The act of detecting and acquiring Reed Solomon blocks on a CDPD channel.

T200   M-ES retransmission timer.

T201   The minimum time between TEI check messages.

T203   The length of time to wait before an M-ES enters sleep mode.

T204   How much time may elapse before TEI notification messages are retransmitted.

T205   The length of time an M-ES will wait before transmitting acknowledgments of received INFO frames on the reverse channel.

TCP (transmission control protocol)   A connection-oriented transport service as defined in RFC793.

TDM (time-division multiplexing)   Provision of multiple channels on a single communications facility by mapping specific users to specific channels via assigned time slots.

TDMA (time division multiple access)   A medium access method which provides multiplexing capability based on time slot assignments as opposed to frequency or contention mechanisms.

TEI (temporary equipment identifier)   A dynamically assigned MDLP address which is associated with only one registered M-ES within any given serving area.

Telex   International teletypewriter service.

TELNET   A standards-based virtual terminal application that is part of the TCP/IP protocol suite.

Temporary equipment identifier   See **TEI.**

TEST   An MDLP message which enables service provider operator-controlled remote loopbacks of an M-ES.

TIA   Telecommunications Industries Association.

Topology database   A database resident in intermediate systems which describes reachable networks and link states.

TP4   Class 4 of the ISO Transport Protocol. Provides connection-oriented transport services, similar to RFC793 TCP.

Transmission burst   The sum of bits transmitted by an M-ES over a CDPD reverse channel between the time a transmitter ramps up and, upon completion of transmission, ramps down.

Transport connection   A connection between communicating end systems by either the TP4 or TCP process.

Transport protocol   A layer 4 protocol that is end-system resident. In the case of CDPD, TCP and TP4 have been defined as the protocols to use for connection-oriented services, while UDP has been assigned for connectionless transport requirements.

Tree   Part of a V.42bis dictionary entry.

Type code   A vendor community protocol identifier. Associated with identifying network-level protocols.

UA (unnumbered acknowledgment)   An affirmative response to an HDLC mode-setting command. Supported by CDPD's MDLP.

UDP (user datagram protocol)   A connectionless transport service as defined by RFC768.

UI (unnumbered information)   Transmitted information that has no sequence numbers associated with it. Used by broadcast and TEI management entities.

Uncorrectable error   A bit error that cannot be compensated for by Reed Solomon processes at the MAC level. Will result in ultimate retransmission by the link level protocol, MDLP.

Undirected hop   The term used to describe a hop initiated by an M-ES with no assistance or intervention by the CDPD network.

V.22   CCITT standard for 1200-bits-per-second transmission.

V.22bis   CCITT standard for 2400-bits-per-second transmission.

V.32   CCITT standard modulation scheme for two-wire full-duplex communications.

V.32bis   CCITT standard modulation scheme for two-wire full-duplex transmission from 4800 to 14,400 bits per second. Backward compatible with V.32.

V.42   An HDLC-based error correcting scheme for modems. Also known as LAPM.

V.42bis   An international data compression algorithm.

V.120   An international rate adaption scheme that is HDLC based.

VLR (visitor location register)   Found within the AMPS system, analogous to CDPD's MSF.

WASI   Wide area service identifier.

Wireless   A mode of communicating via radiofrequency, infrared, or other means where no physical connections are required between communicating end systems. Sometimes referred to as "free space" communications.

WWW (World Wide Web)   A collection of participating databases and protocols used within the Internet that enables hypertext links among various files sharing a common or similar topic.

X OR   A Boolean math operation.

X.121   An international numbering plan for public data networks. Associated with X.25.

X.25   An international standard for communications via public packet switching networks. X.25 provides connection-oriented services via both permanent and switched virtual circuits.

X.400   An international message handling system.

XID (exchange ID)   An HDLC verb which is part of CDPD R1.0 registration procedures, but which has been omitted from the CDPD R1.1 repertoire.

ZAP   An MDLP verb which shuts down a malfunctioning transmitter at an M-ES.

802.x   A series of local area network standards defined by the IEEE.

Bibliography

Bishop, Peter, "Securing Your CDPD Applications and Data," CDPD Forum presentation, 1995.

CCITT, Recommendation V.42bis, 1992.

CCITT, Recommendation X.213, 1992.

CDPD Forum, Recommendation 1.0, July 1993.

CDPD Forum, Recommendation 1.1, January 1995.

Cincinnati Microwave, "A Primer on How CDPD Can Change the Way Your Organization Does Business," July 21, 1995.

Dellecave, Tom, Jr., "Wireless Off the Ground," *Information Week,* May 29, 1995, pp. 33–34.

Doyle, Thomas E., "Wireless Is the Ticket for Police," *Wireless for the Corporate User,* Vol. 3, No. 6, 1994, p. 51.

EIA/TIA, "553 Mobile Station–Land Station Compatibility Specification," 1989.

Fist, Stewart, "Will GSM and D-AMPS Give Way to the CDMA Push," *Australian Communications,* July 1995, pp. 83–88.

Hewlett Packard, "IS-41 MAP Protocol Analysis," *Telecommunications News,* January 1995, pp. 14–15.

Jones, Vincent C., *MAP/TOP Networking,* McGraw-Hill, New York, 1988.

Knightson, Keith G., Terry Knowles, and John Larmouth, *Standards for Open Systems Interconnection,* McGraw-Hill, New York, 1988.

Lee, William C. Y., *Mobile Cellular Telecommunications,* McGraw-Hill, New York, 1995.

Lin, C., and D. J. Costello, Jr., *Error Control Coding: Fundamentals and Applications,* Prentice-Hall, Englewood Cliffs, NJ, 1983.

Macario, Raymond C. V., *Cellular Radio: Principles and Design,* Macmillan, London, 1993.

Mason, Charles F., "CDPD: Getting It Ready for Users," *Wireless for the Corporate User,* Vol. 4, No. 1, 1995, pp. 30–35.

Mason, Charles F., "For Hertz, Speed Is Gold," *Wireless for the Corporate User,* Vol. 4, No. 3, 1995, pp. 12–20.

Odenwalder, Joseph P., *Error Control,* SAMS, Indianapolis, IN, 1985.

Qualcomm, Inc., "CDMA vs. GSM: A Comparison of the Seven C's of Wireless Communications," 1995.

RFC 791, Internet Protocol DARPA Internet Program Protocol Specification, 1981.

RFC 793, Transmission Control Protocol DARPA Internet Program Protocol Specification: A Proposal for Multi-Protocol Transmission of Datagrams Over Point-to-Point Links, 1981.

RFC 941, Addendum to the Network Service Definition Covering Network Layer Addressing, April 1985.

RFC 982, Guidelines for the Specification of the Structure of the Domain Specific Part (DSP) of the ISO Standard NSAP Address, April 1986.

RFC 986, Guidelines for the Use of Internet IP Addresses in the ISO Connectionless-Mode Network Protocol, June 1986.

RFC 1237, Guidelines for OSI NSAP Allocation in the Internet, July 1991.

RFC 1441, Introduction to Version 2 of the Internet Standard Network Management Framework, April 1993.

Slekys, Arunas G., "High Capacity Cellular for Wireless Telephony," *Hughes Network Systems,* October 1993.

Stephens, Terrence G., "Understanding Spread Spectrum Radio," *Communications Design,* October 1995, pp. 51–53.

Wexler, Joanie, "CDPD Struggles to Take Packet Radio Nationwide," *Network World,* October 16, 1995, p. 8.

Winch, Robert G., *Telecommunication Transmission Systems,* McGraw-Hill, New York, 1993.

Index

ACSP (*see* AMPS channel status protocol)
Adjacent channel scan, 132
Adjacent neighbor, 130
Advanced Mobile Phone System (AMPS), 9
AMPS (*see* Advanced Mobile Phone System)
AMPS channel status protocol (ACSP), 122–123
ARN (*see* Authentication random number)
ASN (*see* Authentication sequence number)
Authentication random number (ARN), 115
Authentication sequence number (ASN), 115

BLER (*see* Block error rate)
Block error rate (BLER), 126, 135–136

CD (*see* Collision detection)
CDMA (*see* Coded division multiple access)
CDPD (*see* Cellular digital packet data)
Cell configuration message, 130–131
Cell number, 128
Cells, 9
Cellular digital packet data (CDPD):
  application layer, 42
  applications of, 2–7, 139–171
  authentication, 111–118
  billing, 8
  building blocks of, 23
  channel attributes, 54–60
  data link control layer, 39–40, 53–54
  network layer, 40–41
  physical layer, 37–38
  presentation layer, 42
  profile, 33
  registration, 111–118
  session layer, 41–42
  transport layer, 41

Cellular geographical serving area (CGSA), 28, 63
CGSA (*see* Cellular geographical serving area)
Channel gain, 12
Channel hopping, 11, 20, 119–138
Channel management, 124–126
Channel quality parameter message, 134
Checkpointing, 81
Circuit-mode mobile data intermediate system (CMD-IS), 166
Circuit-mode mobile-end system (CM-ES), 166
Circuit switched CDPD (CS-CDPD), 166
Circuit switched CDPD control protocol (CSCCP), 166
Circuit switched mobile base station (CS-MB), 166
Circuit switched networks, 8
CLSs (*see* Connectionless services)
CM-ES (*see* Circuit-mode mobile-end system)
CMD-IS (*see* Circuit-mode mobile data intermediate system)
Co-channel interference (C/I), 62-63
Coded division multiple access (CDMA), 13
Collision detection (CD), 60
Color codes, 62–64
Compression:
  network-level data compression, 101
  network-level header compression, 101
Configuration timer, 75
Connection-oriented services (COSs), 43
Connectionless services (CLSs), 43
Constant carrier, 56
Continuity indicator, 66
Control channels, 56
Control flag, 60
COSs (*see* Connection-oriented services)